ACS SYMPOSIUM SERIES 774

Agrochemical Discovery
Insect, Weed, and Fungal Control

Don R. Baker, EDITOR
Berkeley Discovery

Noriharu Ken Umetsu, EDITOR
Otsuka Chemical Company

American Chemical Society, Washington, DC

Library of Congress Cataloging-in-Publication Data

Agrochemical discovery : insect, weed, and fungal control / Don R. Baker, editor, Noriharu Ken Umetsu, editor.

p. cm.—(ACS symposium series ; 774)

Includes bibliographical references (p.) and index.

ISBN 0–8412–3724–7

1. Agricultural chemicals.

I. Baker, Don R., 1933–   . Arthur G., 1954–   . II. Umetsu, Noriharu Ken., 1946–   . III. Pan Pacific Conference on Pesticide Science (2nd : 1999 : Honolulu, Hawaii IV. Series.

S585 .A56   2000
632´95—dc21   00–56553

Foreword

THE ACS SYMPOSIUM SERIES was first published in 1974 to provide a mechanism for publishing symposia quickly in book form. The purpose of the series is to publish timely, comprehensive books developed from ACS sponsored symposia based on current scientific research. Occasionally, books are developed from symposia sponsored by other organizations when the topic is of keen interest to the chemistry audience.

Before agreeing to publish a book, the proposed table of contents is reviewed for appropriate and comprehensive coverage and for interest to the audience. Some papers may be excluded in order to better focus the book; others may be added to provide comprehensiveness. When appropriate, overview or introductory chapters are added. Drafts of chapters are peer-reviewed prior to final acceptance or rejection, and manuscripts are prepared in camera-ready format.

As a rule, only original research papers and original review papers are included in the volumes. Verbatim reproductions of previously published papers are not accepted.

ACS BOOKS DEPARTMENT

Contents

Synthesis and Chemistry
(Eiichi Kuwano and Don R. Baker)

Natural Products
(Hiroshi Abe and Horace G. Cutler)

Agricultural Biotechnology
(Hideo Ohkawa and William P. Ridley)

Combinatorial Chemistry
(Joseph G. E. Fenyes and Isao Iwataki)

Mode of Action Studies
(Keiichiro Nishimura and Don R. Baker)

vii

Preface

Agrochemical Discovery: Insect, Weed, and Fungal Control provides a unique current look at the process of discovery in the field of crop protection. Crop protection is undergoing a major restructuring in the methods used for insect, weed, and fungal control. The major advances in medical genetic engineering are finding applications in modifying plants to better be able to provide their own crop protection or enhanced yield and quality. Crops of many types are coming that have resistance to specific highly potent herbicides.

Similarly, crop plants are being developed that can control their specific insect or fungal problems. Because of the development of resistance by many types of plants to various control agents, it is necessary to look for entirely new active types of compounds. Combinatorial chemistry approaches long used in medical discovery are now being applied to find new agrochemical types. In this way hundreds of thousands of new compounds can be made and tested for the proper characteristics for safe and economic use in agriculture.

Additionally, natural products are finding increased use in the development of new means and types of crop control agents. The 2nd Pan-Pacific Conference on Pesticide Science held in October 1999 in Honolulu, Hawaii brought together an international focus on these new developments in crop protection. The ideas presented there provide the foundation for the material covered in this new book.

DON R. BAKER
Berkeley Discovery
15 Muth Drive
Orinda, CA 94563

NORIHARU KEN UMETSU
Otsuka Chemical Company
3– 2–27 Ote-dori, Chuo-ku
Osaka 541–0021, Japan

Chapter 1

Modern Agrochemical Discovery

Don R. Baker[1] and Noriharu Ken Umetsu[2]

[1]Berkeley Discovery, 15 Muth Drive, Orinda, CA 94563
[2]Otsuka Chemical Company, 3–2–27 Ote-dori, Chuo-ku, Osaka 541–0021, Japan

The requirements for modern agrochemicals seem almost impossible to satisfy. The public demands materials that are very safe both to humans and the environment. This requires low toxicity to a wide range of non-target species and at the same time short persistence and low residues in crops and water. Farmers want agrochemicals which are active across a wide range of target species and last long enough in the environment to maintain control for that extended period they require. At the same time the compounds have to be safe to workers in the fields and economical to use. It is with this apparent dichotomy, that the 2nd Pan-Pacific Conference on Pesticide Science was held in Honolulu, Hawaii on October 24-27, 1999. (1,2) This conference was a joint effort between the Agrochemicals Division of the American Chemical Society and the Pesticide Science Society of Japan. This present chapter and book provides an examination of this important field of scientific endeavor.

Introduction

The agrochemical community sits at the center of a major world problem: How do we continue to feed the ever-increasing world population? The ultimate fate of civilization rests on how this problem is worked out. (3,4) For many parts of the world this problem is already reaching critical proportions with billions suffering. Most of the agrochemical production and research communities are not located in these underfed countries. This major problem confronting the world should give encouragement to proceed with the time-consuming, costly research needed for new necessary products and methods of agriculture. Means must continue to be found to continue to find ways to provide safe, economical, and effective products to aid the production of both major and minor crops. (5-9) Fortunately, a few organizations continue to see the importance of providing new better and more effective products for the lesser markets for agrochemical agents and means. We must continue to look for new products which possess greater improvement in mammalian toxicity and environmental safety at lower and lower rates in the environment. (10) For this the heart of any discovery program begins with reliable and efficient assays that predict the ability of a chemical to protect a field crop. (11)

Insect Control

Today we are seeing many kinds of insects that are little affected by a host of insect control agents. Almost every crop known to man has one or more insect predators. Each of these take their toll in the quantity and quality of the food and fiber products which come from our farming efforts. Lepidopterous insects continue to be one of the major predators on our food and fiber crops. Genetically modified crops such as corn and cotton offer new means of controlling Lepidoptera. Here genes of *Bacillus thuringensis* (*Bt*) are expressed in the crop plant. The *Bt* toxin is a small protein which acts as a stomach poison on the insect. As with many past selective insect control agents, resistance is beginning to emerge with *Bt*. This points out the importance of finding new safe, effective insect control methods. Also conservation steps are necessary in order to preserve the effectiveness of compounds or methods where resistance can develop.

In those insects where the life cycle is short there seems to be a propensity for the easy development of resistance to insect control agents. Several generations can take place in a single season. This seems to be the case with mosquitoes. Many places in the world have mosquito populations that are resistant to man's most effective control schemes. Many serious public health

problems, such as malaria, encephalitis, yellow fever, and dengue fever find
mosquitoes as disease vectors. Approximately 40% of the world's population
lives in areas of malaria risk. Here there are 300 to 500 million cases annually
and of these 1.5 to 2.7 million die of malaria each year. Of these deaths, 90%
occur in Africa.

Each year there are 30 million cases of dengue fever with at least 25,000
deaths. As yet there is no vaccine for dengue fever. Its main vector is *Aedes
aegypti* and its subspecies. Recently a predator, misocyclops, a small shrimp-
like creature (also called the one-eyed shrimp) was found in Vietnam and
where it was propagated this mosquito vector was greatly controlled. (12)
Unfortunately, the many areas where mosquitoes are disease vectors are also
places which can ill afford or in many cases have little means to control those
diseases born by mosquitoes.

Another public health menace is Chagas Disease, which has a vector in
the night biting kissing bug. About 16 to 18 million people are infected each
year. Most of the 50,000 deaths each year occur in Mexico, and Central and
South America. Some cases are starting to appear as the kissing bug finds its
way into the southern United States. Because of world travel we are now seeing
the appearance of these diseases in the developed countries of the world. (13)
This points to the continuing need for efforts to find new methods for
controlling insects for both agriculture and public health.

Mode of action studies are important in the development of new insect
control agents which offer sufficient selectivity in toxicology tests. A section of
this current work has chapters which explore this important adjunct to the
development of new insect control agents.

Herbicides

Only the major crops can afford the luxury of modern agrochemical
research. Herbicide research and development into new means of weed control
research is limited to corn, soybeans, rice and perhaps wheat. Other crops offer
only small markets for new products. The large chemical manufacturers are
selling off their minor products which were once used in the smaller
agrochemical markets. Because of registration and re-registration costs many
of the minor products are no longer available. New weed pressures are
developing in many of these areas where older off-patent products are being
used.

Resistance is also becoming a problem with herbicides that have a single
site for their mode of action. Resistance has been quick to develop in many
important weeds by the Acetolactate synthase (ALS) inhibiting sulfonyl ureas

and imidazolinones. (14) In all cases studied, the mechanism of resistance has been due to selection of an altered form of the ALS enzyme. These herbicides are exciting because of their wide use at low concentration and their inherent safety. Management of ALS resistance has been through use of other mode of action herbicides in rotation with the ALS controlling compounds.

Mode of action studies are important to the development of new herbicides. Much has happened in the last few years in understanding how xenobiotic compounds affect biological systems. (15) Herbicides which affect enzymes peculiar to plants often offer toxicity characteristics favorable to further development. (16) The current search for new control agents usually involves mode of action studies to determine which lead compounds or enzyme systems should be explored more extensively.

Plant biotechnology offers important opportunities to improve food production methods. (17) Several major products have been approved for use in several countries. This current work explores this type of effort as one of its major section of chapters.

Fungicides

Fungicides fit alongside insecticides and herbicides in the production of agricultural products. Without modern fungicides many of our agricultural products would be much more expensive to grow. Certain growth conditions predispose many crops to fungal attack. In general terms there are two types of fungicides, protectant fungicides and systemic fungicides. The systemic fungicides are ones where the compound moves through the outer tissues of the plant into the phloem or xylem of the plant. Systemic fungicides usually have only one mode of action in the fungi at a site not usually found in plants or animals. Systemic fungicides are usually subject to resistance development due to the fact that fungi can rapidly reproduce and rapidly adapt to the presence of the fungicide. (18)

Protectant fungicides usually do not penetrate any great distance into the plant. Here they remain on the outer surface of the plant giving protection from fungal attack. Many of the common protectant fungicides such as captan have their mode of action on more than one site in the fungal cell. This multiplicity makes it more difficult for the organism to develop resistance to the fungicide. At the same time, because of the compound's interaction with more than one site, there may be toxicology issues where animal or plant sites are effected.

5

References

1. *Abstracts*, 2nd Pan-Pacific Conference on Pesticide Science, Honolulu, Hawaii; October 24-27, 1999; 61 pages.
2. *Final Program*, 2nd Pan-Pacific Conference on Pesticide Science, Honolulu, Hawaii; October 24-27, 1999; 8 pages.
3. Klasen, W., in *Eighth International Congress of Pesticide Chemistry, Options 2000;* Ragsdale, N. N.; Kearny, P. C. ; Plimmer, J. R. , Eds; ACS Conference Proceedings Series, Washington, D. C., 1995; pp 1-32.
4. Fedoroff, V., "Food for a Hungry World: We Must Find Ways to Increase Agricultural Productivity", *The Chronicle of Higher Education* 1997, *43*, Number 41, pp B4-B5.
5. Baker, D. R.; Fenyes, J. G.; Basarab; G. S.: Hunt, D. A., Eds., *Synthesis and Chemistry of Agrochemicals V*, ACS Symposium Series #686. American Chemical Society, Washington D. C.: 1998, 340 pp.
6. Baker, D. R.; Fenyes, J. G.; Basarab, G. S. Eds., *Synthesis and Chemistry of Agrochemicals IV*, ACS Symposium Series #584. American Chemical Society, Washington D. C.: 1995, 490 pp.
7. Baker, D. R.; Fenyes, J. G.; Steffens, J. J., Eds., *Synthesis and Chemistry of Agrochemicals III*, ACS Symposium Series #504. American Chemical Society, Washington D. C.: 1992, 468 pp.
8. Baker, D. R.; Fenyes, J. G.; Moberg, W. K., Eds., *Synthesis and Chemistry of Agrochemicals II*, ACS Symposium Series #443. American Chemical Society, Washington D. C.: 1991, 609 pp.
9. Baker, D. R.; Fenyes, J. G.; Moberg, W. K.; Cross, B., Eds., *Synthesis and Chemistry of Agrochemicals*, ACS Symposium Series #355. American Chemical Society, Washington D. C.: 1987, 474 pp.
10. Umetsu, N. K.; *Pesticides and Human Health - To Investigate the Safety of Agrochemicals.* Japan Plant Protection Association, 1998 (in Japanese).
11. Basarab, G. S.; Baker, D. R.; Fenyes, J. G. Bioassays in the Discovery Process of Agrochemicals. *Synthesis and Chemistry of Agrochemicals IV,* ACS Symposium Series, #584, American Chemical Society, Washington, D. C. pp. 1-14, (1995).
12. Hewetson, M. , "Dengue Fever Management," Radio National - Earthbeat (Australia); February 20, 1999.
13. Pray, W. S.; *US Pharmacist*, 21 (6): 18, 20-22, 24, 27; 1996 Jacobson Pubublishers.
14. Powles, S. B.; Holtum, J.A.M.; *Herbicide Resistance in Plants Biology and Biochemistry*; CRC Press, Boca Raton, FL,1994, 368 pp.

15. Pike, D. R.; Hagar, A.; McGlamery, M.; *How Herbicides Work*, University of Illinois. Extension, Dept. of Crop Sciences; 1998; Available on the Internet.
16. Schmidt, R. R.: HRAC Classification of Herbicides according to Mode of Action.; *Brighton Crop Protection Conference – Weeds;* pp 1133-1140, 1997
17. *Economics of Innovation and New Technology;* Harwood Academic, The Gordon and Breach Pub. Group, ISSN: 1043-8599; 6 issues per volume.
18. McGrath, M. T.; Zitter, T. A. "Managing Fungicide Resistance for Powdery Mildew, Gummy Stem Blight, and Downey Mildew;" *Capitol Vegetable News;* July 1996; Available on the Internet.

Synthesis and Chemistry

With an ever expanding world population new agrochemicals will continue to play a significant role in producing food crops and fibers. Environmentally safer and more efficacious agrochemicals become increasingly necessary for agricultural production. Development of new agrochemicals begins with the searching for new lead compounds. However, any discovery strategy is difficult to predict and program. Historically most new leads have been discovered accidentally or serendipitously. Today, in the discovery of new agrochemicals there is a major need for successful rational approaches to finding worthwhile new leads and for the ability to synthesize a wide variety of new compounds almost at will.

Success stories of agrochemical discovery, most of which involve new classes of chemistry and optimization of the characteristics necessary for commercialization through the process of synthesis and biological evaluation. This gives courage to synthetic chemists to work for that creative leap, aimed at the discovery of new types of effective agrochemicals.

The following synthesis chapters deal with the discovery of a new herbicide, fungicide and insecticide which have attained commercial importance as useful agrochemicals or are currently being developed for commercial use. Triketone herbicides which inhibit p-hydroxyphenylpyruvate dioxygenase is reviewed by D.L. Lee and colleagues. Cyclopropanecarboxamide fungicides, an inhibitor of melanin biosynthesis is considered by K. Wada and associates. Finally, G.P. Lahm and coworkers review oxadiazine insecticides, a new sodium channel blocker.

Eiichi Kuwano
Laboratory of Pesticide Chemistry
Division of Bioresources and Bioenviornmental Sciences
Graduate School, Kyushu University
Hakozaki 6-10-1, Higashi-ku
Fukuoka 812-8581 JAPAN

Don R. Baker
Berkeley Discovery
15 Muth Drive
Orinda, CA 94563

Chapter 2

The Synthesis and Structure–Activity Relationships of the Triketone HPPD Herbicides

David L. Lee[1], Christopher G. Knudsen, William J. Michaely,
John B. Tarr, Hsiao-Ling Chin, Nhan H. Nguyen, Charles G. Carter,
Thomas H. Cromartie, Byron H. Lake, John M. Shribbs,
Stott Howard, Sean Hanser, and D. Dagarin

Zeneca Ag Products, Western Research Center,
1200 South 47th Street, Richmond, CA 94804
[1]Current email address: leedavidl@aol.com

The 2-benzoylcyclohexane-1,3-diones, the triketones, are a novel class of bleaching herbicides whose mode of action is the inhibition of p-hydroxyphenylpyruvate dioxygenase. The synthesis and structure-activity relationships of this chemical class are discussed.

Targeting the inhibition of p-hydroxyphenylpyruvate dioxygenase (HPPD) as a herbicidal mode of action is a recent, exciting development in the pesticide industry. Most, if not all, of the major agrochemical companies now have or have had research projects in the area. Discovery of HPPD as an unique and viable herbicide target was first made by ZENECA Agrochemicals in the eighties as part of our investigation of herbicidal 2-benzoyl-1,3-cyclohexanediones (*1*). Elucidation of HPPD as the unique site of action of these compounds was first made in mammalian systems (2,3). This site of action was subsequently confirmed in plants (4-8), and a hypothesis for the minimum substructure necessary for the inhibition of HPPD was also developed (3,9). The commercial compound sulcotrione (10), a post-emergent broadleaf herbicide for use in corn in Europe, eventually resulted from those efforts along with the development compound mesotrione for the pre- and post-emergent broadleaf corn market in the US. An additional unexpected benefit from this research was the discovery of the pharmaceutical use of these triketones for the treatment of tyrosinaemia Type I (11). The structure-activity relationships (SAR) of these molecules were elucidated. The SAR of the benzoyl moiety of the triketones were presented in a previous paper (12) and this paper will detail some of our work in the elucidation of the SAR of the dione moiety.

Structure-Activity Relationships

For analysis of the SAR of the triketones the molecule can be conveniently separated into two parts: the benzoyl moiety and the dione moiety (Figure 1). Each part can be examined separately as the two parts appear to play a distinct and different role in the overall expression of the herbicidal activity, and in general, are independent of each other. As such, the activity of the triketones can possibly be described by a Free-Wilson type summation of a separate quantitative descriptor for the dione and the benzoyl group.

Structure Activity Relationships of Benzoyl Moiety

The primary result of adding electron-withdrawing substituents to the phenyl ring is to increase the acidity of the molecule and/or to increase the intrinsic affinity of the molecule for the HPPD enzyme. This has been previously reviewed (12), and it was found that the herbicidal activity was best correlated with the acidity of the triketone molecule, which was affected by the ortho and para substituents of the phenyl ring. Presumably increased acidity is important for transport and translocation within the plant, and it also seems to increase affinity for the enzyme active site. Addition of certain meta substituents also had the result of increasing the affinity of the molecule for the enzyme. Whether causal or not is open to speculation, but we did observe a strong correlation between acidity and herbicidal activity.

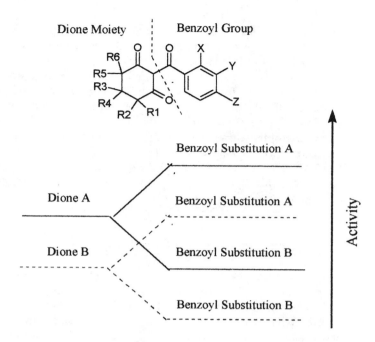

Dione Moiety | Benzoyl Group

Figure 1. Segmentation of the SAR of the triketone molecule

Structure Activity Relationships of Dione Ring

The functional effect of adding substituents to the dione ring is to block site(s) of metabolism of the molecule by plants. This results in greater herbicidal activity as the plants have greater difficulty in detoxifying the molecule. The principal route of metabolism of the triketone molecule in plants is hydroxylation in the 4-position of the 1,3-cyclohexanedione ring as shown in Figure 2. If the 4-position is blocked by substituents, hydroxylation then occurs in the chemically equivalent 6-position of the dione ring.

This potentiation in activity is more evident in grass species than in broadleaf weed species and is dramatically illustrated in Table 1 with a series of progressively methylated dione analogs. As the potential sites of hydroxylation are sequentially blocked (Compounds 2, 3, 4 & 5), one can easily observe a progressive increase in grass activity. Evidence that this increased herbicidal activity was not due to increased binding affinity for the maize HPPD enzyme was provided by comparing the binding affinities for compounds 6 and 7 shown in Table 2. Both the unsubstituted cyclohexanedione analog as well as its 4,4-

dimethyl counterpart have similar binding affinities, yet the 4,4-dimethyl analog has significantly greater herbicidal activity towards grasses.

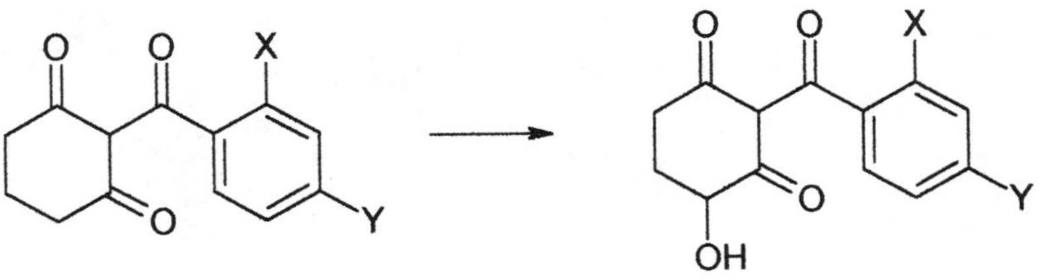

Figure 2. Principal route of metabolism of the triketones in plants

Table 1. Effect of Dione Substitution on Activity

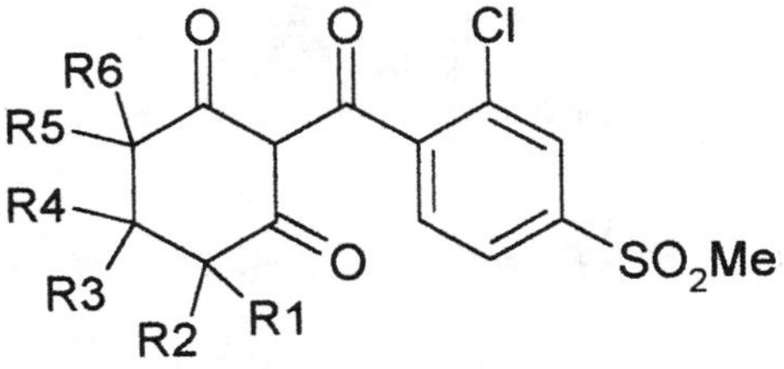

Cpd #	R1	R2	R3	R4	R5	R6	LD_{50} GR $(g/ha)^a$
1	H	H	H	H	H	H	143
2	Me	Me	H	H	H	H	75
3	Me	Me	H	H	Me	H	59
4	Me	Me	H	H	Me	Me	23
5	Me	Me	C=O		Me	Me	9

[a]Rate in grams per hectare required to obtain an average weed control rating of 50% on six different grass weed species (DIGSA, ECHCG, LOLRI, SETFA, SETVI, & SORHA) in a pre-emergent application.

Table 2. Effect of Dione Substitution on Binding Affinity for Maize HPPD Enzyme

Cpd #	R1	R2	LD$_{50}$ GR (g/ha)[a]	HPPDI IC$_{50}$ (nM)
6	H	H	89	27
7	Me	Me	11	28

[a]Rate in grams per hectare required to obtain an average weed control rating of 50% on six different grass weed species (DIGSA, ECHCG, LOLRI, SETFA, SETVI, & SORHA) in a pre-emergent application.

Effect of Substitution with Groups Other than Methyl

The effect on herbicidal activity of substitution of the cyclohexanedione ring with groups other than methyl such as ethyl, alkoxy, alkylthio, alkylsulfinyl, alkylsulfonyl, halo, imino, oximino, and propargyl have also been examined (*13-16*). Besides blocking potential sites of metabolism on the cyclohexanedione ring, factors such as metabolism of the functionality itself, size, and the log P of the substitutent were expected to be important. In general, these analogs were less active than the corresponding methyl substituted compounds. Though there were changes in activity, crop selectivity, and weed spectrum with these substituent changes, there were no discernible structure-activity patterns associated with this myriad of compounds with the lone exception of a minimum log P requirement. A broad range of functional groups are tolerated on the cyclohexanedione ring; however, if the phenyl ring is already substituted with a polar group, such as methylsulfonyl, substitution of yet another methylsulfonyl group on the cyclohexanedione ring is not tolerated (compare compounds 1 and 8 in Table 3). Note that methylsulfonyl substitution on the cyclohexanedione ring is easily tolerated on an analog where there are no other polar groups present (compare compounds 10 and 11 in Table 3).

Table 3. Effect of Functional Group Substitution on Activity

Cpd #	X	Y	R1	R2	LD_{50} GR (g/ha)[a]
1	Cl	SO$_2$Me	H	H	143
8	Cl	SO$_2$Me	SO$_2$Me	H	>>280
9	Cl	SO$_2$Me	OMe	H	344
10	NO$_2$	CF$_3$	H	H	17
11	NO$_2$	CF$_3$	SO$_2$Me	H	38

[a]Rate in grams per hectare required to obtain an average weed control rating of 50% on six different grass weed species (DIGSA, ECHCG, LOLRI, SETFA, SETVI, & SORHA) in a pre-emergent application.

Propesticides

As previously noted (12), triketones can be viewed as vinylogous benzoic acids. Accordingly, they can be functionalized to vinylogous acid chlorides (17), amides (18,19), esters (20,21), and thioesters (22-24) as shown in Figure 3. Such precursors can all regenerate the triketone active ingredient through hydrolysis. As can be expected, the level of herbicidal activity expressed by these derivatives were best correlated with their rate of hydrolysis, and thus, their primary utility was for the time-release of the parent triketone. Factors such as the pH of the medium, water solubility of the compound, and steric hindrance surrounding the vinylogous carbonyl system of these propesticides were found to be important for their activity

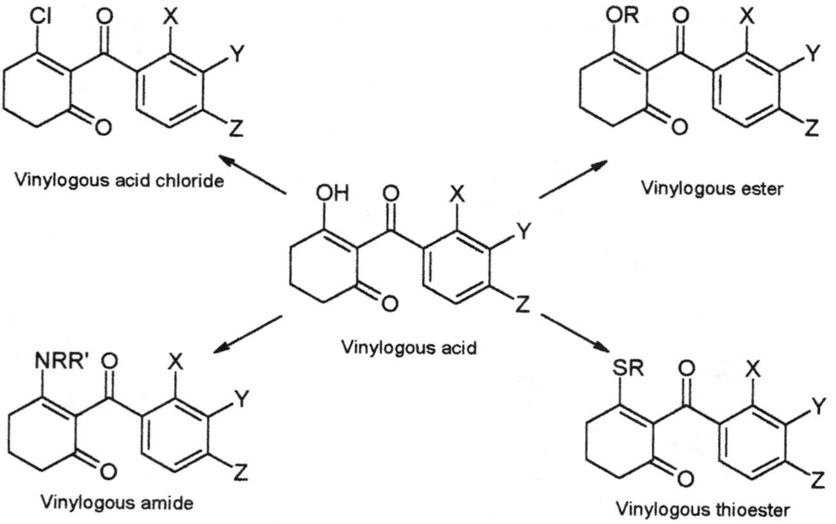

14

Figure 3. Classes of propesticides of the triketones

Another class of propesticides is the 9-oxotetrahydroxanthenones. These compounds are, in essence, 2-(2-hydroxybenzoyl)-1,3-cyclohexanediones that are in intramolecular equilibrium with their tetrahydroxanthenone propesticidal form (Figure 4). Presumably this conversion occurs via intramolecular attack of the ortho hydroxyl group on one of the carbonyl groups on the cyclohexanedione ring to form a cyclic hemiketal. Elimination of water then affords the tetrahydroxanthenone. We first became suspicious that these tetrahydroxanthenones may indeed be intramolecularly masked 2-benzoyl-1,3-cyclohexanediones when purified samples of these compounds still exhibited the bleaching herbicidal activity characteristic of the triketone herbicides. Since these tetrahydroxanthenones had been prepared from herbicidally active ortho-chloro or ortho-nitrobenzoyl-1,3-cyclohexanediones, we had originally attributed the herbicidal activity observed with crude samples of these compounds to residual amounts of the parent triketone. However, highly purified samples of these xanthenones, which were devoid of parent triketone, were still found to be herbicidally active. Since these tetrahydoxanthenones did not possess the requisite 2-benzoylethenol substructure required of an HPPD inhibitor (3), we were forced to the hypothesis that these compounds were in equilibrium with the active triketones. Proof that this equilibria existed was obtained by heating a mixture of a tetrahydroxanthenone with methyl iodide in acetone-water. Complete conversion to the 2-methoxybenzoyl-1,3-cyclohexanedione was obtained.

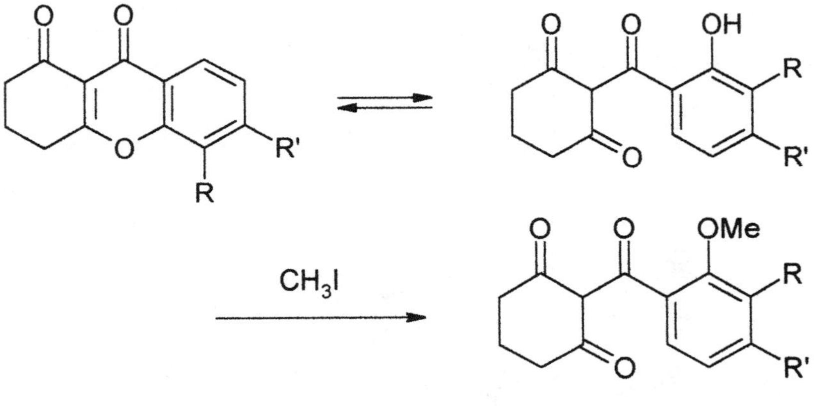

Figure 4. Proposed equilibrium between tetrahydroxanthenone and triketone

Synthesis

The synthesis of the triketones was accomplished using a two-step procedure (*25*) as shown in Figure 5. The first step involved the reaction of a benzoyl chloride with the cyclohexane-1,3-dione in the presence of a base such as triethylamine to afford the intermediate enol ester. The enol ester was then rearranged to the triketone via a catalytic amount of acetone cyanohydrin. The mono, di, and trisubstituted cyclohexane-1,3-diones were prepared via reaction of a suitably substituted acrylate with a methyl ketone in a solvent such as toluene in the presence of a molar equivalent of a base such as sodium ethoxide (*13-16*)(Figure 6). 4,4,6,6-Tetramethylcyclohexane-1,3-dione was prepared from 4,4,6-trimethylcyclohexane-1,3-dione through the formation of the dianion followed by quenching with one equivalent of methyl iodide as shown in Figure 7.

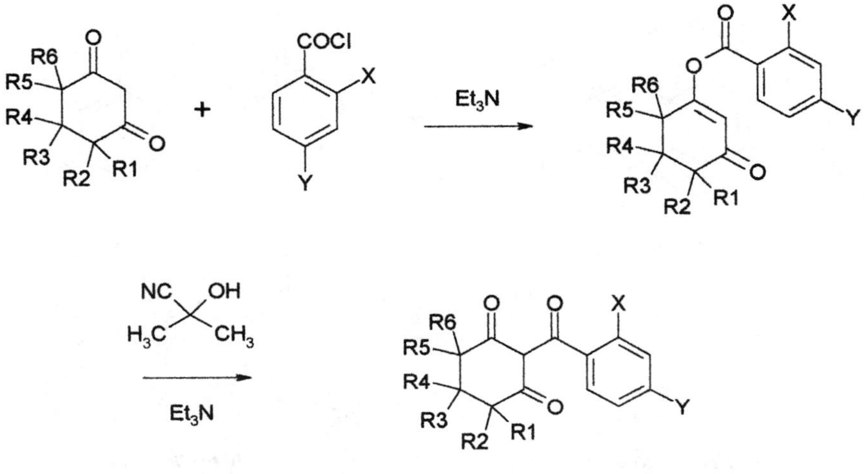

Figure 5. Synthesis of the triketones via enol ester rearrangement

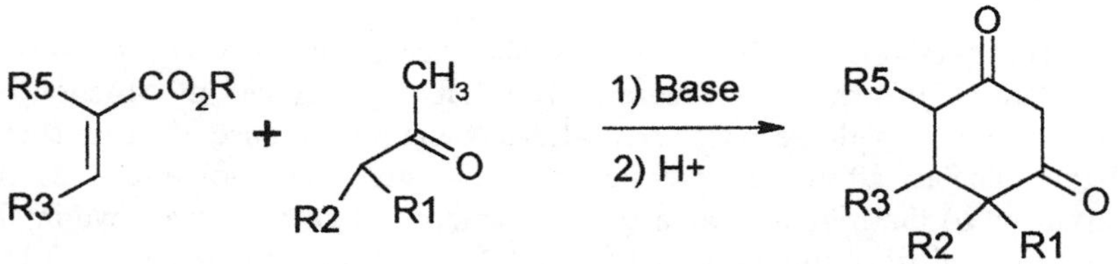

Figure 6. Synthesis of Substituted Cyclohexane-1,3-diones

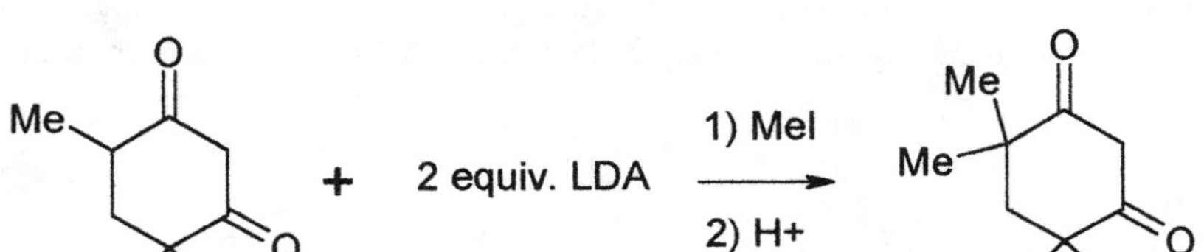

Figure 7. Synthesis of 4,4,6,6-tetramethylcyclohexane-1,3-dione

Summary

The structure-activity relationships of the triketone class of HPPD inhibiting herbicides can be separated into two distinct and independent components reflecting different parts of the molecule: the benzoyl moiety and the cyclohexane-1,3-dione moiety. For the benzoyl portion, substituents on the phenyl ring, which increase the acidity of the molecule, correlated well with increased activity and generally with the intrinsic affinity of the molecule for the HPPD enzyme. Adding electron-withdrawing groups to the ortho and/or para positions had the most effect in increasing acidity. The addition of meta substituents had the effect of increasing the affinity of the molecule for the enzyme. For the cyclohexane-1,3-dione portion, the functional effect of adding substituents to the dione ring was to block site(s) of metabolism of the molecule by plants. This resulted in greater herbicidal activity as detoxification of the molecule was slowed since the principal route of metabolism of the triketone molecule in plants is hydroxylation in the 4-position of the 1,3-cyclohexanedione ring. This potentiation of activity was more evident in grass species than in broadleaf weed species

References

1. Michaely, W. J. and Kratz, G. W. Certain 2-(2-substitutedbenzoyl)-1,3-cyclohexanediones. EP 0 135 191, 1986.

2. Ellis, M. K., Whitfield, A. C., Gowans, L. A., Auton, T. R., Provan, W. M., Lock, E. A., and Smith, L. L., Inhibition of 4-hydroxyphenylpyruvate dioxygenase by 2-(2-nitro-4-trifluoromethylbenzoyl)-cyclohexane-1,3-dione and 2-(2-chloro-4-methylsulfonylbenzoyl)-cyclohexane-1,3-dione. *Toxicol. Appl. Pharmacol.*, **133** (1995) 12-19.

3. Lee, D. L., Prisbylla, M. P., Cromartie, T. H., Dagarin, D. P., Howard, S. H., Provan, W. M., Ellis, M. K., Fraser, T., and Mutter, L. C., The discovery and structural requirements of inhibitors of p-hydroxyphenylpyruvate dioxygenase. *Weed Science*, **45** (1997).

4. Prisbylla, M. P., Onisko, B. C., Shribbs, J. M., Adams, D. O., Liu, Y., Ellis, M. K., Hawkes, T. R., and Mutter, L. C., The novel mechanism of action of the herbicidal triketones. *Proc. British Crop Prot. Conf. - Weeds*, **2** (1993) 731-738.

5. Prisbylla, M. P., Lee, D. L., Cromartie, T. H., Dagarin, D. P., and Howard, S. W., Mode of action studies on structurally related triketone HPPD

inhibitors. *Inter. Chem. Cong. Of Pacific Basin Soc.Book of Abs.*, **1** (1995) 94.

6. Schultz, A. Ort, O., Beyer, P., and Kleinig., SC-0051, a 2-benzoylcyclohexane-1,3-dione bleaching herbicide, is a potent inhibitor of the enzyme p-hydroxyphenylpyruvate dioxygenase. *FEBS Lett.*, **318** (1993) 162-166.

7. Secor, J., Inhibition of barnyardgrass 4-hydroxyphenylpyruvate dioxygenase by sulcotrione.*Plant Physiol.*, **106** (1994) 1429-1433.

8. Barta, I. C. and Boger, P., Benzoylcyclohexanedione herbicides are strong inhibitors of purified p-hydroxyphenylpyruvic acid dioxygenase of maize. *Pestic. Sci.*, **45** (1995) 286-287.

9. Lee, D. L., Mutter, L. C., Dagarin, D. P., Howard, S. W., and Fraser, T., Structure-activity relationships of HPPD herbicides: Benzoylphenols and β-diketones. *Inter. Chem. Cong. Of Pacific Basin Soc.Book of Abs.*, **1** (1995) 91.

10. Beraud, M., Claument, J., and Montury, A., ICIA0051, A new herbicide for the control of annual weeds in maize. *Proc. British Crop Prot. Conf. - Weeds*, **1** (1993) 51-56.

11. Ellis, M. K., Lindstedt, S. T., Lock, E. A., Markstedt, M. E. H., Mutter, L. C., and Prisbylla, M. P., Use of 2-(2-nitro-4-trifluoromethylbenzoyl)-1,3-cyclohexanedione in the treatment of tyrosinaemia and pharmaceutical compositions. WO 93/00080, 1992.

12. Lee, D. L., Knudsen, C. G., Michaely, W. J., Chin, H., Nguyen, N. H., Carter, C. G., Cromartie, T. H., Lake, B. H., Shribbs, J. M., Fraser, T., The structure-activity relationships of the triketone class of HPPD herbicides. *Pestic. Sci.*, 54 (1998) 377-384.

13. Lee, D. L., Certain 2-(2-substituted benzoyl)-4-(substituted oxy or substituted thio)-1,3-cyclohexanediones. US Patent 4 783 213, 1988.

14. Chin, H. M., Certain 2-(2-substituted benzoyl)-4-(substituted)-1,3-cyclohexanediones. US Patent 4 781 751, 1988.

15. Knudsen, C. G., Certain 2-(2-substituted benzoyl)-4-(substituted imino, oximino, or carbonyl)-1,3-cyclohexanediones. US Patent 4 838 932, 1989.

16. Nguyen, N. H., Certain 2-(2'-substituted benzoyl)-4-propargyl-1,3-cyclohexanedione herbicides. US Patent 4 997 473, 1991.

17. Knudsen, C. G., 3-Chloro-2-(2'-substituted benzoyl)-cyclohex-2-enone intermediate compounds. US Patent 4 837 352, 1989.

18. Knudsen, C. G., Certain substitued 3-amino-2-benzoylcyclohex-2-enones. US Patent 4 775 411, 1988.

19. Knudsen, C. G., Certain substituted bis(2-benzoyl-3-oxo-cyclohexenyl)diamines. US Patent 5 173 105, 1992.

20. Knudsen, C. G. and Michaely, W. J., Certain substituted 3-(substituted oxy)-2-benzoyl-cyclohex-2-enones. US patent 4 918 236, 1990.

21. Knudsen, C. G. and Michaely, W. J., Certain substituted 3-(substituted oxy)-2-benzoyl-cyclohex-2-enones. US patent 4 957 540, 1990.

22. Knudsen, C. G., Certain 3-(substituted thio)-2-benzoyl-cyclohex-2-enones. US Patent 4 762 551, 1988.

23. Knudsen, C. G., Method of controlling undersirable vegetation utilizing certain 3-(substituted thio)-2-benzoyl-cyclohex-2-enones. US Patent 4 854 966, 1989.

24. Knudsen, C. G., Certain substituted bis(2-benzoyl-3-oxo-cyclohexenyl) thioglycols. US Patent 5 152 826, 1992.

25. Heather, J. B. and Milano, P. D., Process for the production of acylated 1,3-dicarbonyl compounds. US Patent 4 695 673, 1987.

Chapter 3

Evolution of the Sodium Channel Blocking Insecticides: The Discovery of Indoxacarb

George P. Lahm, Stephen F. McCann, Charles R. Harrison, Thomas M. Stevenson, and Rafael Shapiro

DuPont Agricultural Products, Stine-Haskell Research Center, Building 300, Newark, DE 19714

Indoxacarb is a new broad spectrum insecticide that acts via blocking of the sodium channel, a new mechanism of action for insect control. It combines a favorable environmental profile with low mammalian toxicity and excellent crop protection. The discovery of indoxacarb evolved through the discovery and evaluation of several new classes of sodium channel blockers. The history of this process is described.

Introduction

The pyrazoline insecticides were discovered by Wellinga, Mulder and van Daalen (*1-6*) in the early 70's and since that time there has been considerable attention focused on this class (*7-12*). Analogs such as PH 60-42 were found to have good activity on both lepidoptera and coleoptera at rates competitive with commercial standards (*13*). Jacobson found that 4-carbomethoxy pyrazolines such as RH3421 were an exceptionally active class with improved soil residual characteristics (*14, 15*). However, we believe issues associated with environmental persistence and potential for bioaccumulation as well as non-target toxicity toward birds, fish and beneficial insects interfered with successful commercialization of the general class of pyrazoline insecticides (*16, 17*).

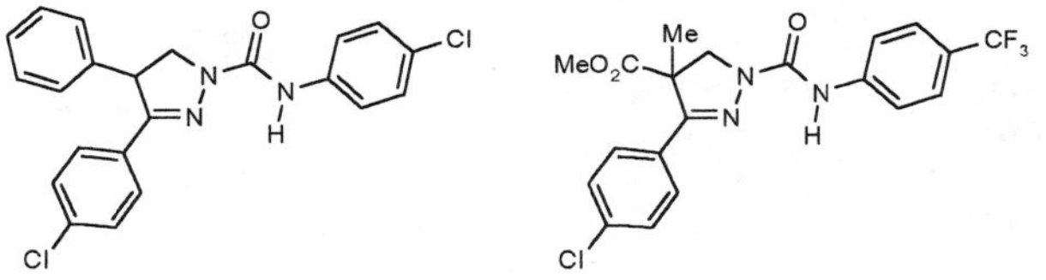

PH 60-42
Philips-Duphar

RH3421
Rohm and Haas

Figure 1: Philips Duphar and Rohm and Haas Pyrazolines

The finding by Salgado (*18*) that these compounds act through blocking of the sodium channel, a mechanism similar to that of local anesthetics, was significant as it suggested the potential for a new mode of action insecticide. And while effort was expended toward optimization of substitution patterns around the pyrazoline ring no advances were made toward modification or replacement of the pyrazoline nucleus and no complete solution to the array of toxicity and environmental problems was identified. Our objective then became to balance the complex dynamic of optimizing insecticidal activity against that of reducing non-target detrimental attributes. We would like to report on our efforts in this area culminating in the discovery of the new insecticide indoxacarb.

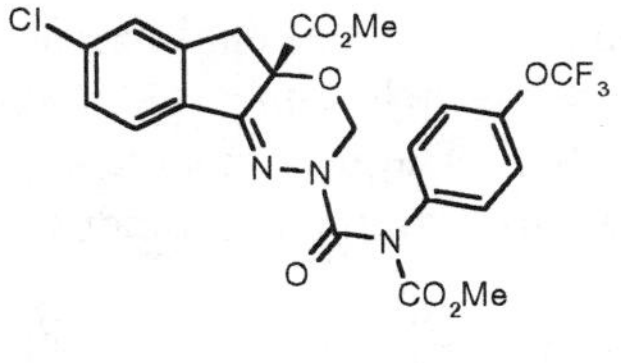

Figure 2: Indoxacarb

The discovery of indoxacarb was an evolutionary process that progressed through development of a variety of new chemical/structural classes. From the lead compound of Philips-Duphar (I) we sequentially investigated carboxamides (II), indazoles (III), semicarbazones (IV), pyridazines (V) and oxadiazines (VI). We believe structures I-VI are united by a common mode of action, i.e. blocking of the sodium channel, as they share common symptomology and similar structure activity patterns, and perhaps most importantly structures I and VI have been shown to act by this mechanism. Fortuitously, each of these chemical classes also contributed stepwise insight into the ultimate indoxacarb discovery.

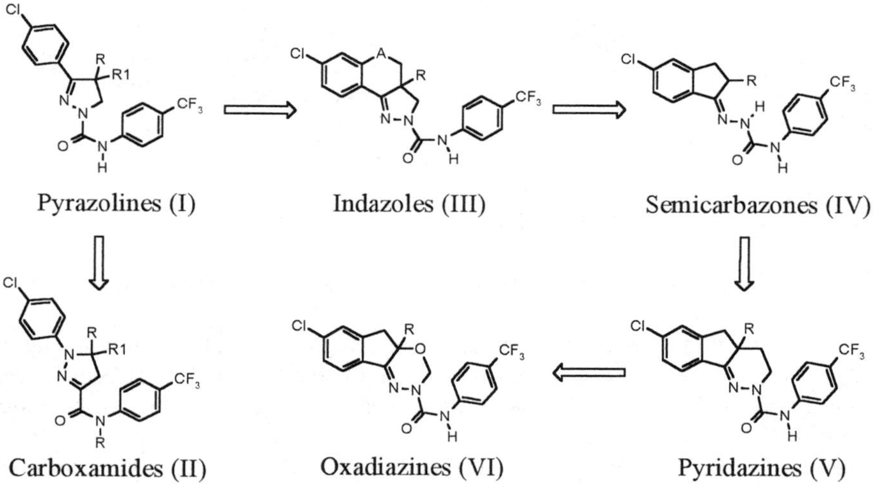
Pyrazolines (I)
Indazoles (III)
Semicarbazones (IV)
Carboxamides (II)
Oxadiazines (VI)
Pyridazines (V)

Figure 3: Evolution of the Structure of the Sodium Channel Blocking Insecticides

Pyrazoline Carboxamides

Initially we investigated pyrazoline carboxamides of formula II wherein the pyrazoline functionality was held intact, while the 1-nitrogen and 3-carbon atoms were inverted (*19-21*). These compounds show striking overlap with molecular models of known pyrazoline insecticides.

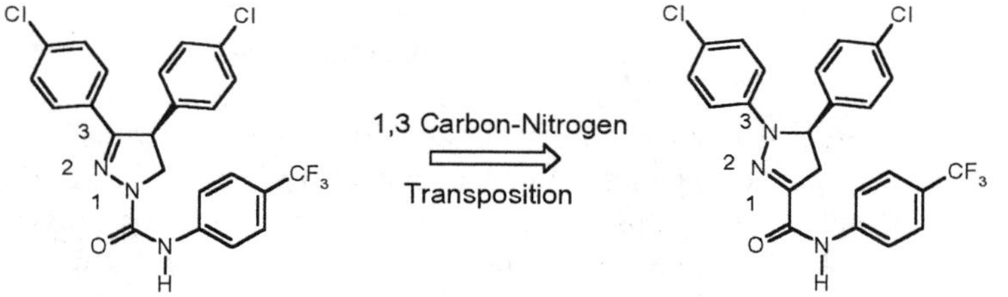
1,3 Carbon-Nitrogen
Transposition

Figure 4: 1,3 Carbon-Nitrogen Transposition of Pyrazoline Insecticides

Hydrazonyl chlorides were prepared through a modification of the Japp-Klinngeman reaction and subsequently converted to 3-carbomethoxy pyrazolines via 3+2 cycloaddition with a wide variety of olefins. The ester group was converted to the target carboxamides via standard methods (Figure 5).

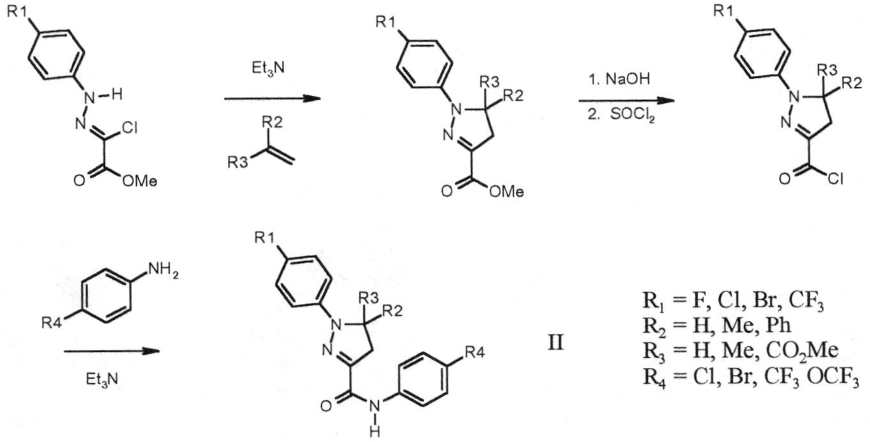

R$_1$ = F, Cl, Br, CF$_3$
R$_2$ = H, Me, Ph
R$_3$ = H, Me, CO$_2$Me
R$_4$ = Cl, Br, CF$_3$ OCF$_3$

Figure 5: Preparation of 1-Phenyl-3-Carboxamide Pyrazolines

Insecticidal activity of these compounds tracked closely with known structure activity relationships of traditional pyrazoline insecticides. This provided an indication that pyrazoline modification was possible and offered potential for alteration of compound attributes. However, for these compounds greenhouse and field tests indicated a lack of sufficient residual insect control, and therefore further development was not pursued.

Indazoles

The X-ray crystal structure of pyrazoline I-a shows this compound adopts a planar π configuration held intact by an intramolecular hydrogen bond at N-2, with substituents in the 4-position of the pyrazoline out of this plane (Figure 6). We reasoned that we could effectively lock this conformation by joining the 4-position of the pyrazoline ring with the ortho position of the 3-aryl group.

Figure 6: Conformationally Restrained Pyrazolines

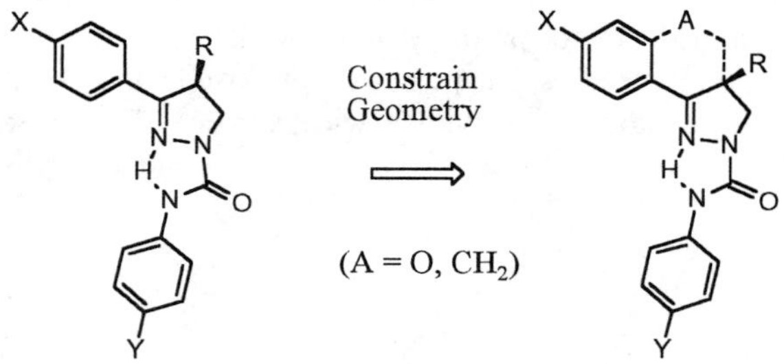

General procedures for the preparation of these analogs followed methods similar to previously reported syntheses of 4,4-disubstituted pyrazolines. Indazoles containing angular aryl groups were prepared from 2-aryl tetralones (*22, 23*). Sequential treatment of the 2-aryl tetralone with formaldehyde and base followed by methanesulfonyl chloride afforded the intermediate mesylate. Conversion to indazoles III-b was accomplished by treatment with hydrazine in refluxing n-butanol followed by reaction with aryl isocyanates.

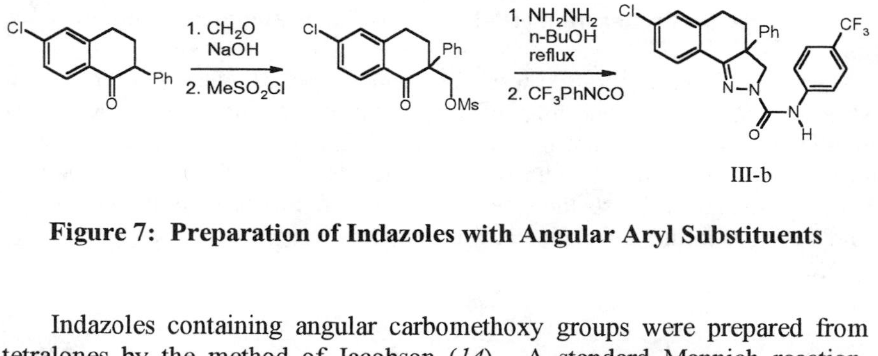

Figure 7: Preparation of Indazoles with Angular Aryl Substituents

Indazoles containing angular carbomethoxy groups were prepared from tetralones by the method of Jacobson (*14*). A standard Mannich reaction, followed by sequential reaction with hydrazine and aryl isocyanate afforded the intermediate indazoles (III-c) lacking an angular substituent. Treatment with two equivalents of lithium diisopropylamide, followed by methyl chloroformate afforded angular carbomethoxy indazoles of formula III-d.

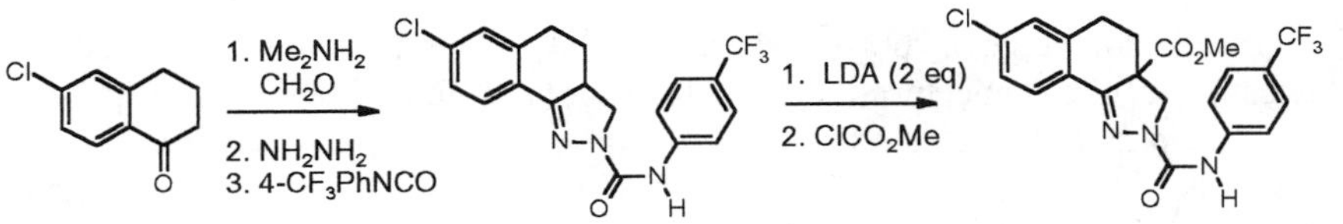

Figure 8: Preparation of Indazoles with Angular Carbomethoxy Substituents

Synthesis of oxyindazoles of formula III-e involved as the key step an intramolecular cycloaddition of a phosphorylsemicarbazide with an internal olefin. Acidic cleavage of the phosphoryl group followed by reaction with aryl isocyanate afforded the target oxyindazoles. Generally incorporation of oxygen increased insecticidal activity and comparative data of indazole/oxyindazole pairs (Table 1) typically showed greater activity for the oxyindazoles.

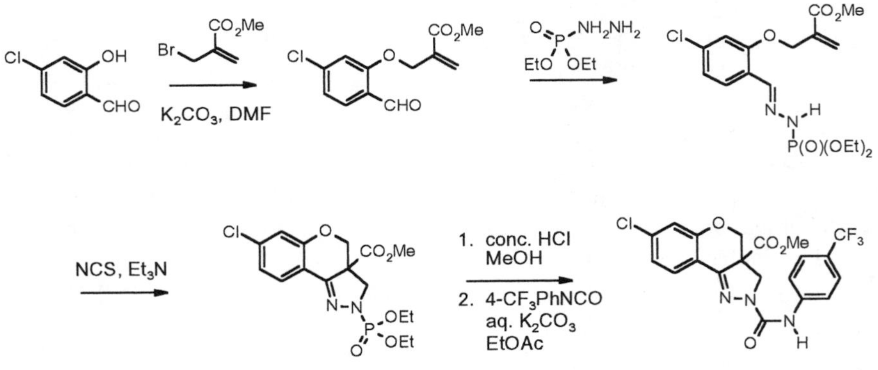

Figure 9: Preparation of Oxyindazoles

Insecticidal data is presented in Table 1. The LEC80's represent the lowest effective concentration at which greater than 80% mortality was observed and indicate good activity at rates competitive with commercial standards. We also separated enantiomers for two of the indazoles and confirmed that activity resided in the S-enantiomer, consistent with the report of Bosum-Dybus and Neh for a 4-phenylpyrazoline (*24*).

Table 1: Insecticidal Activity – Indazoles and Oxyindazoles

--------------- LEC80 (ppm) -------------

X	R	TBW	FAW	SCRW	BW
CH_2	4-Cl-Ph	50	50	250	50
CH_2	CO_2Me	250	10	2.5	10
O	4-Cl-Ph	2.5	2.5	100	50
O	CO_2Me	2.5	2.5	2.5	10

Tobacco budworm (TBW, Heliothis virescens), Fall Armyworm (FAW, Spodoptera frugiperda), Southern Corn Rootworm (SCRW, Diabrotica undecimpunctata howardi), Boll Weevil (BW, Anthonomus grandis grandis), Colorado Potato Beetle (CPB, Leptinotarsa decemlineata)

Semicarbazones

2-Aryl semicarbazones (IV) were postulated to be good pyrazoline mimics. It was apparent from molecular models that the combined changes of removal of the 5-methylene with simultaneous addition of a bridging group would place the 2-substituent of the semicarbazone in a position of space similar to the 4-position of the pyrazoline and we expected the same planar π configuration held intact by an intramolecular hydrogen bond between the aniline NH and C=N carbazone.

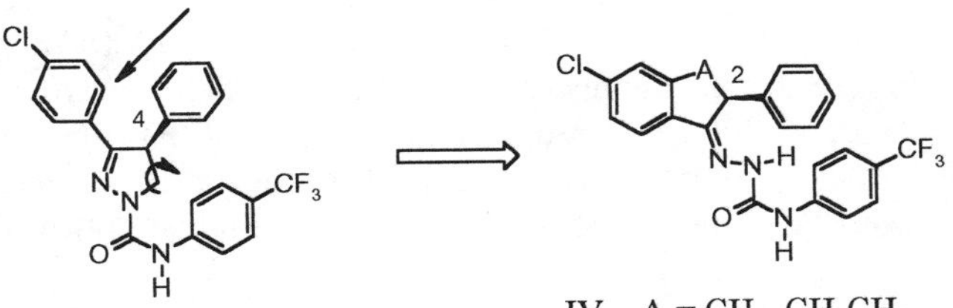

IV   A = CH$_2$, CH$_2$CH$_2$

Figure 10: Combining Strategies of Conformational Locking with Ring Opening

2-Aryl semicarbazones containing methylene and ethylene bridging groups were prepared from the corresponding indanone and tetralone intermediates (25). As a class, these compounds were found to possess good levels of insecticidal activity, although compounds containing a methylene spacer (A = CH$_2$) were found to be significantly more active than those containing an ethylene unit (A = CH$_2$CH$_2$). This was in contrast with findings observed for indazoles where insect activity followed the trend A = CH$_2$CH$_2$ > CH$_2$.

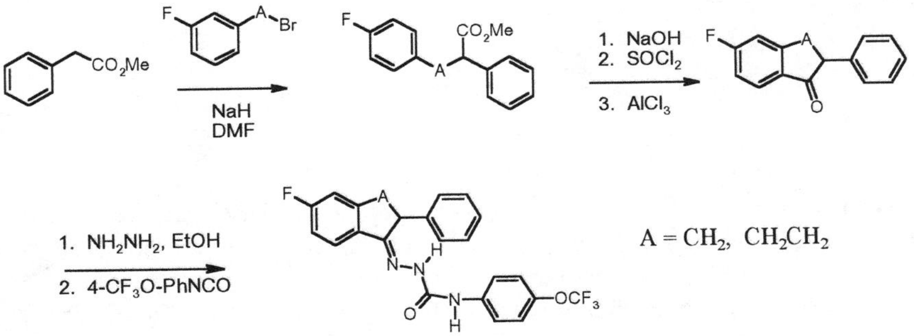

A = CH$_2$, CH$_2$CH$_2$

Figure 11: Preparation of Semicarbazones

With semicarbazones we also observed differences among compound attributes including insect spectrum, soil residual and potential for bioaccumulation. Insecticidal data for three analogs is summarized in Table 2. Compound IV-a showed very good lepidopteran activity but suffered from long soil persistence and the apparent potential to bioaccumulate. Conversely the bromo analog, IV-b, showed a reduced soil half life and an improved metabolic profile while maintaining very good lepidopteran activity. Alkyl substituted semicarbazones such as IV-c displayed very good activity on beetles, but were consistently weaker on lepidoptera. Against Colorado potato beetle II-c provided outstanding crop protection at field use rates of 7.5-20 gm/ha. This compound also displayed both a short soil half life and favorable metabolic profile. These results encouraged the search for broader spectrum compounds with these favorable attributes of mammalian and environmental safety.

Table 2: Insecticidal Activity of Semicarbazones

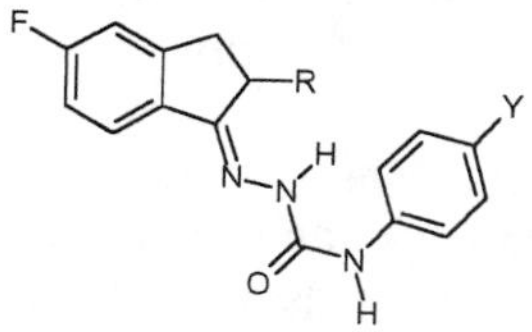

| Cmpd | R | Y | Lepidoptera | | Coleoptera | | |
			FAW	HV	SCRW	BW	CPB
IV-a	4-F-Ph	OCF3	< 1	< 1	50	50	--
IV-b	4-F-Ph	Br	5	5	50	10	--
IV-c	Me	CF3	250	--	< 2.5	10	< 1

-------------------- LEC80 (ppm)--------------------

Pyridazines

The geometric requirements for insecticidal activity suggested that ring size could influence activity. Molecular models indicated reasonable overlap between indazoles, semicarbazones and the corresponding six membered pyridazine analogs, particularly with the expected influence of the internal hydrogen bond on conformation (Figure 12).

Indazoles   Semicarbazones   Pyridazines

Figure 12: Evolution from Indazoles to Pyridazines

Pyridazine V-a (*26*) was prepared from 2-carbomethoxy chloroindanone (Figure 13). Alkylation with dibromoethane afforded the intermediate bromoethyl indanone as well as significant amounts of O-alkylated product. Treatment of the bromoethyl indanone with hydrazine and then 4-trifluoromethoxyphenyl isocyanate afforded V-a. This compound was found to have exceptional activity on lepidoptera and demonstrated excellent crop protection at field use rates of 10-20 gm/ha on fall armyworm. Unfortunately, we also observed unfavorable soil characteristics that precluded further development.

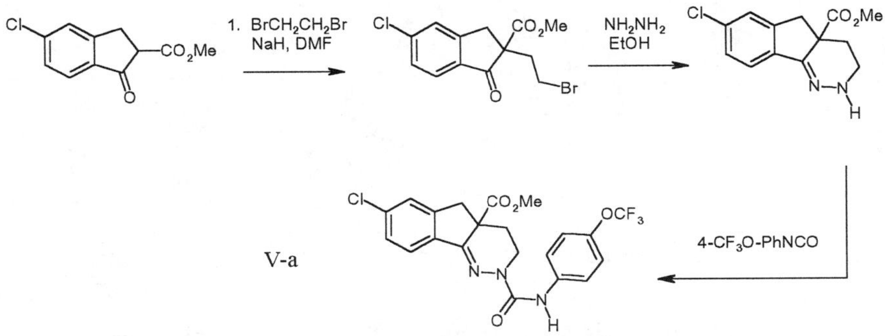

Figure 13: Preparation of Pyridazines

Defining the Pharmacophore of Sodium Channel Blocking Insecticides

With the combined information on chemical classes I-V we postulated a model pharmacophore of the sodium channel blocking insecticides (Figure 14). Substituents X are located in a well defined region of space and optimally positioned through their relationship with the extended pharmacophore through and including the aniline ring. The range of X groups is relatively broad, but most preferred groups are selected from small, lipophilic, electron withdrawing substituents, including halogen, haloalkyl and haloalkoxy. The Y substituent is preferred in the para position and also selected from a small group of electron withdrawing substituents, optimally including chloro, bromo, trifluoromethyl and trifluoromethoxy.

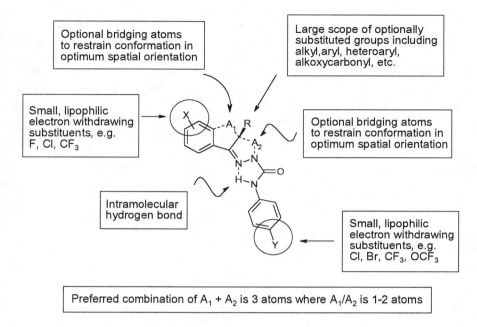

Preferred combination of A_1 + A_2 is 3 atoms where A_1/A_2 is 1-2 atoms

Figure 14. Proposed Pharmacopore of Sodium Channel Blocking Insecticides

X-ray structure analysis indicates the free NH group forms an intramolecular hydrogen bond with nitrogen at N-2. We believe the role is to assist the compound in adopting a planar configuration held intact through this intramolecular bond which creates a preferred stereochemical orientation. It seems likely that this configuration is adopted at the active site and is supported through the observation that the absence of this intramolecular bond eliminates activity and all structure-activity relationships of the different classes of sodium channel blocking insecticides can be rationalized through this orientation.

Perhaps the most flexibility in structural variation can be found in the R substituent. A broad range of active R groups have been identified which include alkyl, aryl, alkoxycarbonyl, heteroaryl and various nitrogen linked substituents. However, the influence of the R substituent can vary widely, effects can be specific to insect spectrum, and R substituent potency is observed to vary among the chemical classes. Therefore, while structure-activity relationships of the R group are not completely defined it is fair to conclude that this group plays a vital role in levels of activity and spectrum.

The bridging groups A_1 and A_2 play a key role in locking the pharmacophore in a preferred orientation and can be viewed as constraining the X/Y termini in an optimal configuration. While the A_1/A_2 groups can be present or absent, a preferred combination of A_1 and A_2 appears to comprise a total of 3 atoms selected from carbon, oxygen and nitrogen, where each A_1/A_2 is 1-2 atoms. We believe this combination constrains the intact pharmacophore in its most preferred configuration and the allows for optimal orientation of substituent groups at X, R and Y.

Oxadiazines

Oxadiazine targets evolved from a threefold perspective: (1) we possessed a good working knowledge of the pharmacophore; (2) issues associated with environmental persistence and bioaccummulation remained; and (3) synthetic accessibility of key pyridazine targets continued to be a problem. We noted that oxadiazines fit the preferred pharmacophore quite well and expected these compounds to be inherently more fragile. With this we hoped that persistence and accumulation problems could be mitigated. Furthermore, we envisioned that synthetic accessibility would be straightforward provided a key cyclization step with formaldehyde and a 2-hydroxy semicarbazone could be achieved.

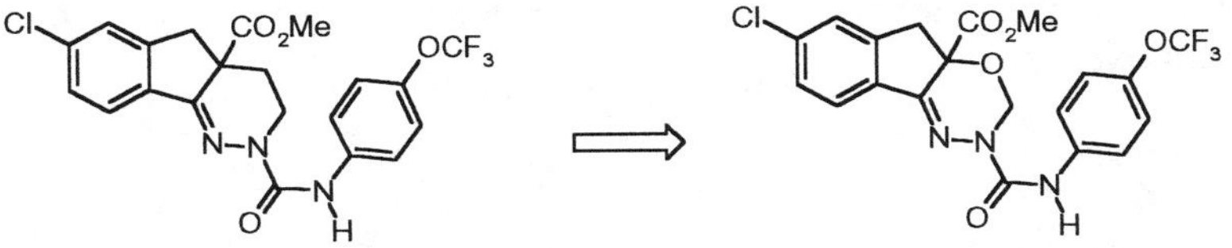

Figure 15: Replacement of a Pyridazine Carbon with Oxygen. Evolution of Pyridazines to Oxadiazines.

Synthesis of the target oxadiazines proceeded as outlined in Figure 16 (27). 2-Carbomethoxy chloroindanone was oxidized with m-chloroperbenzoic acid, condensed with hydrazine and coupled with 4-trifluoromethoxyphenylisocyanate to provide semicarbazone VI-a. The key ring closure was achieved by condensation of VI-a with formaldehyde in the presence of a catalytic amount of p-toluenesulfonic acid providing the oxadiazine target VI-b. Compound VI-b possessed all of the desirable attributes we were seeking. It showed excellent insecticidal activity, was found to have a short soil half life and displayed a favorable toxicological profile both acute and cumulative.

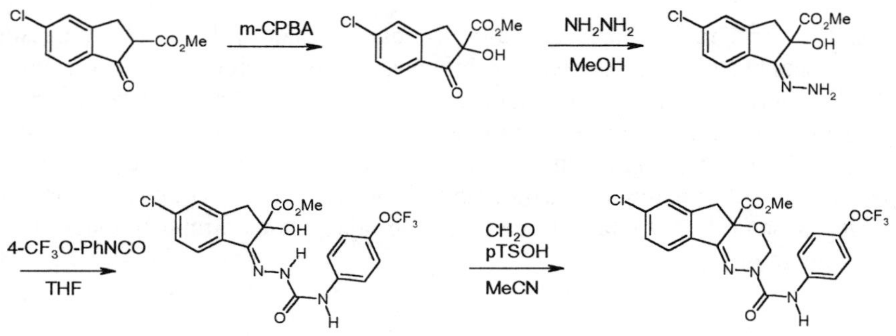

32

Figure 16: Preparation of Oxadiazines

Derivatization of the free NH had the advantage of maintaining excellent insecticidal efficacy while further improving environmental and mammalian safety. Studies by Wing established that the N-carbomethoxy compounds are in fact pro-insecticides and that VI-c (DPX-JW062) is itself only weakly active in the blocking action of sodium channels. However, it appears to be rapidly metabolized by insects to the potent sodium channel blocker VI-b which is responsible for insecticidal activity (*28*). We believe that other routes of breakdown either through metabolic detoxification in mammals or microbial breakdown in the environment occur preferentially to provide the wide safety margins we observe.

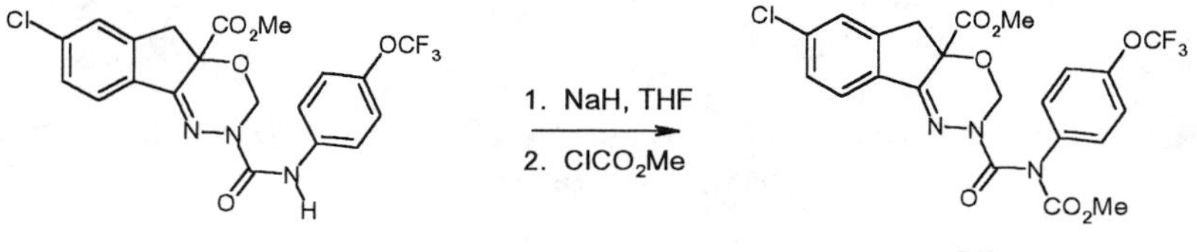

Figure 17: N-Derivatization of Oxadiazines

We confirmed our expectation that the (S)-enantiomer would be principally responsible for activity through asymmetric synthesis of the respective enantiomers. Asymmetric oxidation of 2-carbomethoxy chloroindanone with the osmium based dihydroxylation reagents AD-mix-α or

AD-mix-β, followed by recrystallization from isopropyl acetate, provided separately the enantiomeric pair of hydroxy indanones each greater than 99% ee. These were converted to the respective (S) and (R) oxadiazines by methods previously discussed for the synthesis of the racemic material VI-b. Enantiomeric identification was accomplished through X-ray analysis of a derivative and determined that essentially all activity was retained in the (S)-enantiomer.

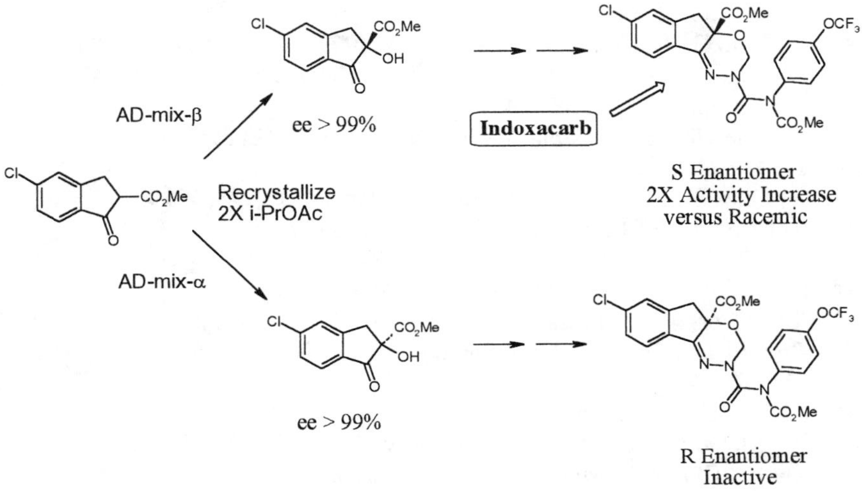

AD-mix-β

ee > 99%

Indoxacarb

S Enantiomer
2X Activity Increase
versus Racemic

Recrystallize
2X i-PrOAc

AD-mix-α

ee > 99%

R Enantiomer
Inactive

Figure 18: Effects of Chirality: Synthesis of Indoxacarb

The (S)-enantiomer, indoxacarb, was found to be highly efficacious on a wide spectrum of insects with excellent crop protection. It combines the favorable attributes of rapid breakdown in the environment, low mammalian toxicity both acute and cumulative, and safety toward birds, fish and beneficial insects. It is, therefore, a significant breakthrough for insect control that fills the need for a new mode of action insecticide. We are confident that it will become a valuable tool for the future in crop protection.

References

1. Mulder, R., Wellinga, K., Van Daalen, J. J., *Naturwissenschaften*, **1975**, 62, 531-532.
2. Wellinga, K., Mulder, R., U.S. 3,991,073, **1976**.

34

3. Wellinga, K., Grosscurt, A. C., Van Hes, R., *J. Agric. Food Chem.*, **1977**, 25, 987-92.
4. Van Hes, R, Wellinga., K., Grosscurt, A. C., *J. Agric. Food Chem.*, **1978**, 26, 915-18.
5. Sirrenberg, W., Klauke, E., Hamman, I., Stendel, W., German Patent DE 2,700,289, **1978**.
6. Van Daalen, J. J., Mulder, R., U.S. 4,174,393, **1979**.
7. (a) Sirrenberg, W., Klauke, E., Hamman, I., Stendel, W., German Patent DE 2,700,288, **1978**.
8. (b) Hunt, J., D.; European Patent 4,733, **1979**.
9. (c) Nakajima, Y., Tsugeno, M., Ishii, S., Hatanaka, M., Hirose, M., Kudo, M., European Patent 58,424, **1982**.
10. (d) Van Hes, R. Grosscurt, A. C., Mebius, E. J.; European Patent 21,506, **1981**.
11. (e) Giles, D. P., Willis, R. J., European Patent 113,213, **1984.**
12. (i) Lahm, G. P., European Patent 300,692, **1989.**
13. Van Daalen, J. J., Mulder, R., U.S. 4,070,365, **1979**.
14. (a) Jacobson, R. M., U.S. 4,663,341, **1987**.
15. (b) Jacobson, R. M., *Recent Advances in the Chemistry of Insect Control*, Crombie, L., E.; The Royal Society of Chemistry: London, **1989**; pp 206-211.
16. (a) Scheele, B., *Chemosphere*, **1980**, 9, 483-494.
17. (b) Meier G. A., Silverman, R., Ray, P. S., Cullen, T. G., Ali, S. F., Marek, F. L., Webster, C. A., *Synthesis and Chemistry of Agrochemicals III*, Baker, D. R., Fenyes, J. G., Steffens, J. J.; American Chemical Society: Washington, DC, **1992**; pp 313-326.
18. Salgado, V. L., *Pestic. Sci.*, **1990**, 28, 389-411.
19. (a) Stevenson, T. M., World Patent 88/05046, **1988**.
20. (b) Stevenson, T. M., World Patent 88/06583, **1988**.
21. (c) Stevenson, T. M., World Patent 89/05300, **1989**.
22. (a) Lahm, G. P., European Patent 286,346, **1988**.
23. (b) Harrison, C. R., Lahm, G. P., Shapiro, R., European Patent 386,892, **1990**.
24. Bosum-Dybus, A., Neh, H., *Liebigs Ann. Chem.* **1991**, 823-825.
25. Daub, J. P., Lahm, G. P., Marlin, B. S., European Patent 377,304, **1990**.
26. Harrison, C. R., Lahm, G. P., Stevenson, T. M., World Patent 91/117983, **1991**.
27. Annis, G. D., Barnette, W. E., McCann, S. F., Wing, K. D., World Patent 92/11249, **1992**.
28. Wing, K. D., Schnee, M. D., Sacher M., Connair, M., *Arch. Insect Biochem. Phys.*, **1998**, 37, 91-103.

Chapter 4

Synthesis and Fungicidal Properties of Cyclopropanecarboxamides

Katsuaki Wada[1], Gerd Hänßler[2], Shinzo Kagabu[3], Udo Kraatz[2], Yoshio Kurahashi[1], and Haruko Sawada[1]

[1]Yuki Research Center, Nihon Bayer Agrochem K. K.,
Yuki, Ibaraki 307–0001, Japan
[2]Landwirtschaftszentrum Monheim, Bayer AG, D–40789 Monheim, Germany
[3]Department of Chemistry, Faculty of Education,
Gifu University, Gifu 501–1193, Japan

Carpropamid (WIN®) is a newly-launched fungicide for controlling rice blast. This compound belongs to a new chemical class of cyclopropanecarboxamides which has been jointly developed by Nihon Bayer Agrochem K.K. and Bayer AG. Carpropamid is classified as a melanin biosynthesis inhibitor (MBI). Compared with other MBIs already on the market, carpropamid has a completely new mode of action, i.e., the inhibition of the dehydration of scytalone to 1,3,8-tri-hydroxynaphthalene or of vermelone to 1,8-dihydroxy-naphthalene in the melanin biosynthesis pathway of *Pyricularia oryzae*.

Introduction

Carpropamid is a new fungicide offering protection against *Pyricularia oryzae*, rice blast. This compound belongs to a new class of fungicides, the

36

cyclopropanecarboxamides, and has a completely new mode of action in contrast to conventional rice blasticides. Carpropamid exhibits excellent protective and systemic efficacy over an extremely long period. These favourable characteristics led us to develop this compound for nursery box applications in controlling both leaf and panicle blast.

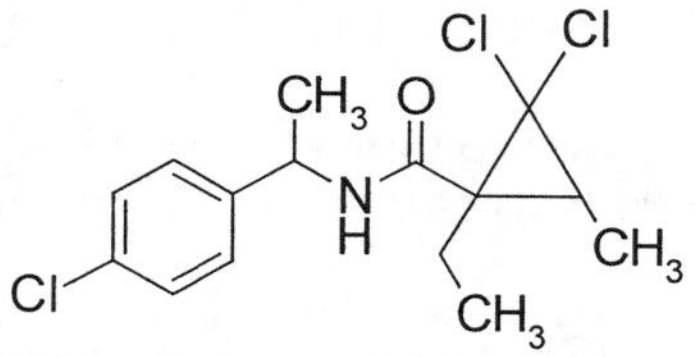

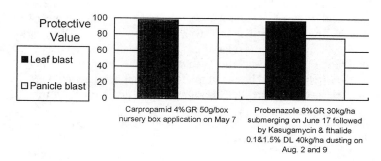

Chemical name :(1*RS*,3*SR*)-2,2-dichloro-*N*-[1-(4-chlorophenyl)ethyl]-1-
ethyl-3-methylcyclopropanecarboxamide
Trade name :WIN (Launched 1998)
Common name : Carpropamid
Code No. : KTU 3616, NTN 33853

Figure 1. Carpropamid: Structure and Nomenclature

Field location: Sendai Miyagi, Japan.
Crop and variety: Rice (Sasanishiki).
Transplanting: May 9, 1994.
Plot size: 960m².
Assessment: Leaf blast, July 31, Panicle blast, Aug. 30.
Damage degree in the untreated plot: Leaf blast (2.5), Panicle blast (1.2)

Figure 2. Efficacy of Carpropamid against Leaf and Panicle Blast in the Field

In Japan, transplanting of nursery plants to the paddy field by transplanting machine is the major method used in rice cultivation. The nursery box application has two advantages. The first is a reduction in manual labour, the second is the decreased environmental risk through lower application rates of the active ingredient. As shown in Figure 2, carpropamid affords excellent efficacy against both leaf and panicle blast following a single application.

History of the Discovery

In 1972, researchers at Shell applied for a patent in which 2,2-dichloro-3,3-dimethylcyclopropanecarboxylic acid, WL 28325, was described as a potent blasticide(*1*). In 1977, Cartwright reported this acid induced phytoalexins, momilactone A and B, when penetration of rice blast started(*2*). This interesting information led us to explore dichlorocyclopropane chemistry, starting in the early 80s.

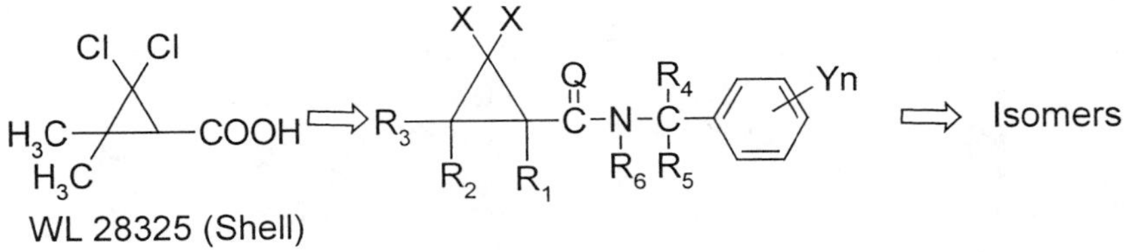

WL 28325 (Shell)

Figure 3. Research Focus on Cyclopropanes

The evolution of our chemistry is depicted in Figure. 3. Until the mid-80s, we concerned ourselves with the chemistry of dichlorocyclopropanecarboxylic acid itself and closely related compounds. Then we found that the α-methylbenzylcyclopropanecarboxamides were an interesting group to be investigated for blasticidal activity, and intensive research was conducted in this area(*3, 4*). The third period is concerned with the stereoisomers which we have been studying until now(*4-6*).

As shown in Table I, compound **1**, WL 28325, showed good performance in greenhouse tests, however, in field trials the efficacy of this compound was unsatisfactory. These results suggest that this molecule is rather volatile and too hydrophilic. Therefore we decided to examine more lipophilic compounds, for example the esters **2** and **3**, and the "higher alkyl" acids **4** and **5**. Then, we replaced the carboxylic acid group with other functionality, e.g. compounds **6-9**. Of particular note is that alcohol **7** and its derivatives showed the same level of activity compared to compound **1**. However, all of the derivatives synthesized did not satisfy the criteria for further development. Simple amides from alkyl or aromatic amines were also synthesized at the same time. Unfortunately, the activity of compounds **10-12** were relatively poor.

Table I. Effect of Substituents on Blasticidal Activity

Compound	R_1	R_2	R_3	X	Activity
1	CH_3	CH_3	H	CO_2H	4
2	CH_3	CH_3	H	$CO_2C_2H_5$	4
3	CH_3	CH_3	H	$CO_2C_6H_5$	3
4	CH_3	CH_3	CH_3	CO_2H	3
5	C_2H_5	C_2H_5	H	CO_2H	2
6	CH_3	CH_3	H	CN	2
7	CH_3	CH_3	H	CH_2OH	4
8	CH_3	CH_3	H	$CH_2OCOC_6H_5$	4
9	CH_3	CH_3	H	$CH_2OCONHC_6H_5$	4
10	CH_3	CH_3	H	$CONH_2$	3
11	CH_3	CH_3	H	$CONHCH_3$	1
12	CH_3	CH_3	H	$CONHC_6H_5$	0

Foliar spray application: at 200ppm.
Inoculation: 1day after treatment.
Evaluation: 7days after inoculation; 5: no lesion, 4:<5% lesion, 3:<20% lesion, 2:<30% lesion, 1:<40% lesion, 0:>40% lesion.

Surprisingly, however, towards the end of this period, we observed a somewhat different biological performance with α-methylbenzylamide compared to the compounds listed in Table I. In particular the residual activity was superior to former cyclopropanes, suggesting new characteristics for these type of compounds. As a result a new synthesis project concerning α-methylbenzylamides started. Blastcidal activity was observed only with a limited number of substituents. α-Methylbenzylamides showed moderate systemic efficacy in greenhouse tests, while some of them showed outstanding performance in field trials. Representative compounds are listed in Table II. On the basis of the field trial data, carpropamid, compound **15**, was selected for development as a commercial fungicide.

Table II. Representative α-Methylbenzylamides

Compound	R_1	R_2	R_3	Yn	Blasticidal activity
13	CH_3	CH_3	CH_3	H	3 - 4
14	CH_3	CH_3	CH_3	4-Cl	5
15	C_2H_5	CH_3	H	4-Cl	5
16	C_3H_7-i	H	H	4-Cl	5
17	CH_3	CH_3	CH_3	4-Br	5

Foliar spray application: at 200ppm.

Inoculation: 1day after treatment.

Evaluation: 7days after inoculation; 5: no lesion, 4:<5% lesion, 3:<20% lesion, 2:<30% lesion, 1:<40% lesion, 0:>40% lesion.

Carpropamid Stereoisomers

Compound **17** shown in Table II, which contains two chiral centers, was selected for the study on the influence of the stereochemistry of the amine component.

Reaction of 2,2-dichloro-1,3,3-trimethylcyclopropanecarboxylic acid chloride with (R)-(+)- or (S)-(-)-4-bromo-α-methylbenzylamines, gave two sets of isomers as shown in Figure. 4.

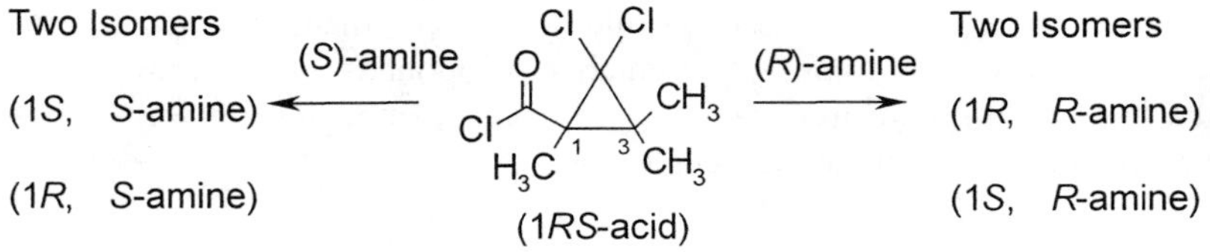

Figure 4. Preparation of Compound 17 using Optically Active Amines

The isomers from (R)-(+)-amine showed + optical rotation, and the isomers from (S)-(-)-amine showed - optical rotation. In this case, separation of the isomers was achieved by chromatography on silica gel eluting with *n*-hexane and ethyl acetate (4:1), the first eluted fraction was called as isomer A and the second eluted fraction as isomer B. Thus, 4 isomers [A+ , B+, A-, B-] were distinguished. The test results indicated that the compounds possessing (R) configuration at the α-methylbenzyl position was far superior in efficacy against rice blast to that of (S) configuration, as shown in Table III.

Table III. Compounds 17 Isomers and Isomer Pairs against Rice Blast

A+	A-	B+	B-	$[\alpha]D$ in EtOH	Efficacy	Melanin synthesis Inhibition
isomer ratio
25	25	25	25	-------	++ - +++	++
50	50	0	0	-------	+++	Not tested
0	0	50	50	-------	+	Not tested
50	0	50	0	+61.2 (c=1.07)	+++	+++
0	50	0	50	- 61.9 (c=1.04)	+	-
100	0	0	0	+64.2 (c=1.13)	+++	+++
0	100	0	0	- 63.9 (c=1.19)	+	-
0	0	100	0	+61.3 (c=1.09)	++	+
0	0	0	100	- 62.6 (c=1.31)	+	-

Foliar spray application: at 50ppm.

Inoculation: 1day after treatment.

Evaluation: 7days after inoculation; +++: no lesion, ++:<5% lesion, +<20% lesion, ->30% lesion.

Melanin inhibition: evaluated by colony color change in agar plate; +++:high, ++:medium, +: weak, -: no change

Structure-activity relationships concerning the cyclopropane configurations were further investigated by using carpropamid. Theoretically there exists $2^3 = 8$ stereoisomers because of three asymmetric centers of carpropamid, however the cyclopropanation of (*E*)-olefine with dichlorocarbene occurs with retention of stereochemistry, affording carproamid as a mixture of only 4 stereoisomers.

As with compound **17**, reaction of (1*RS*,3*SR*)-2,2-dichloro-1-ethyl-3-methylcyclopropanecarboxylic acid chloride with (*R*)-(+)- and (*S*)-(-)-4-chloro-α-methylbenzylamines, respectively, gave two isomers. Again, separation of the isomers was achieved by chromatography on silica gel using with *n*-hexane and ethyl acetate (4:1) as an eluant, and the first eluted fraction was called as isomer A and the second eluted fraction as isomer B. Thus, (*R*)-(+)-amine A+ and B+ isomers, and (*S*)-(-)-amine A- and B- isomers were separated respectively. The absolute configuration of these isomers were determined by X-ray analysis(7).

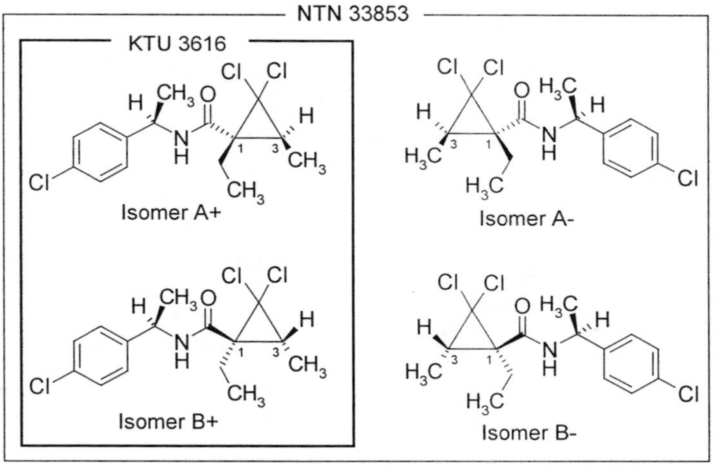

Isomer	Absolute configuration	Efficacy	Melting point
A+	(1*R*, 3*S R*-amine)	++	158 °C
A -	(1*S*, 3*R S*-amine)	+	158 °C
B+	(1*S*, 3*R R*-amine)	+++	162 °C
B -	(1*R*, 3*S S*-amine)	+	162 °C

Efficacy: Summary of rice blast foliar application screening results at 250ppm, 100ppm.

Figure 5.   Absolute Configuration of the Carproamid Stereoisomers

Figure. 5 indicates that the compound possessing (R)-configuration in the amine component leads to superior biological activity against rice blast, with the best results obtained for the isomer B+ (1S, 3R).

Having identified the superiority among isomers we considered it a challenge to incorporate the most potent isomer in enriched form in the active ingredient. This resulted in a switch from NTN33853 (4 isomers) to KTU3616 (2 isomers) which we prepared from the (R)-(+)-enriched amine. Instead of using the optically active amine we also attempted to enrich the most active isomer by an intramolecular diastereoselective reaction as shown in Figure. 6.

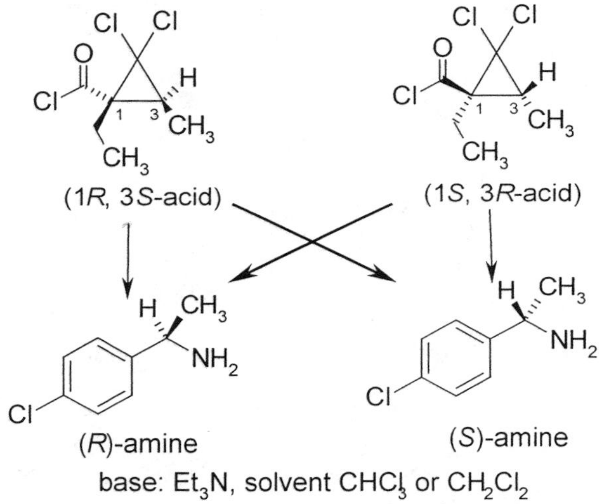

base: Et$_3$N, solvent CHCl$_3$ or CH$_2$Cl$_2$

Isomers ratio at	- 60°C	20°C	60°C
A+(1R, 3S R-amine)	8%	20%	25%
A- (1S, 3R S-amine)	8%	20%	25%
B+(1S, 3R R-amine)	42%	30%	25%
B- (1R, 3S S-amine)	42%	30%	25%

Figure 6. Intra-Molecular Disteleoselectivity of Amidation Step

During investigation of the cyclopropane chemistry, we found that the reaction of (1RS, 3SR)-2,2-dichloro-1-ethyl-3-methylcyclopropanecarboxylic acid chloride with racemic 4-chloro-α-methylbenzylamine gave one enantiomer pair as a predominant form. This diasteroselectivity probably arose from a steric interaction between the ethyl group on the cyclopropane ring and the methyl at the α-position of the amine. At lower temperature, for example at -60°C, the desired isomer pair was formed in excess (~6:1) over the minor pair, as shown in Figure. 6. This crude product (A+ : A- : B+ : B- = 8 : 8 : 42 : 42) was easily purified to afford B+ : B- = 50 : 50 by recrystallization. Thus, we were able to compare two compositions as shown in Table IV.

Table IV.   Efficacy and Physical Properties of Carpropamid Isomer Pairs

A+	A-	B+	B-	Foliar application	Submerged application	Water solubility	Melting point
	Isomers ratio						
50	0	50	0	+++	+++	4.0 mg/L	152°C
0	0	50	50	+++	++	0.3 mg/L	202°C

It was expected that similar level of blasticidal activity should be obtained for the two compositions. The test results, however, indicated that the blasticidal activity of the two compositions were not equal, especially with regard to systemic activity. It is probably due to different physical properties between the two compositions, i.e., the B+/B- racemate forms an extremely stable crystal lattice. Water solubility is the most essential factor regarding systemic action. Therefore we selected the composition for the development from the (R)-(+)-enriched amine.

(R)-(+)-4-chloro-α-methylbenzylamine was prepared in sufficient quantity and optical purity(8). After establishment of this optical resolution procedure, we were able to significantly accelerate the development process.

Mode of Action

Carpropamid is classified as a melanin biosynthesis inhibitor (MBI). Compared with other MBIs already on the market, carpropamid has a completely new mode of action: the inhibition of dehydration of scytalone to 1,3,8-trihydroxynaphthalene or of vermelone to 1,8-dihydroxynaphthalene in the melanin biosynthesis pathway(9).

This inhibition site was further investigated and reported by Kurahashi et al(*10*). Recently the results on X-ray crystal structure analysis for the enzyme, scytalone dehydratase, complexed with carpropamid, was reported(*11*).

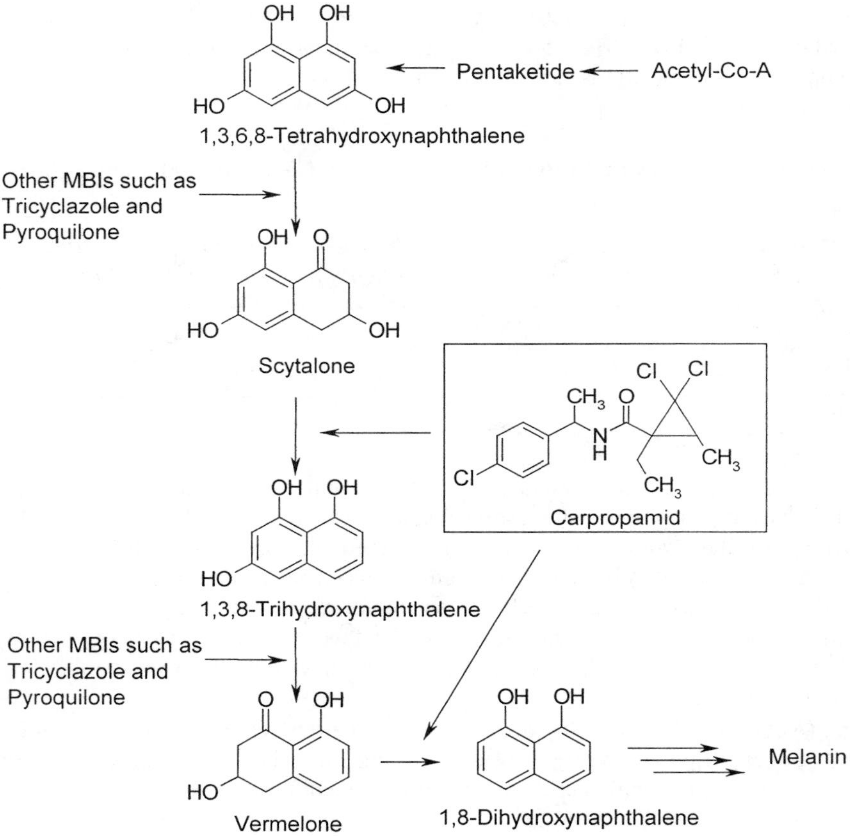

Figure 7. Inhibition Sites of Carpropamid in the Melanin Biosynthesis

Very recently, a secondary effect of carpropamid, i.e., the enhancement of phytoalexin synthesis has been confirmed(*12*). Pre-treatment of carpropamid enhanced the accumulation of phytoalexins, momilacton A and sakuranetin, in rice leaves subsequently inoculated with the blast pathogen (*Pyricularia oryzae*), the similar result was obtained with WL 28325.

Conclusion

Carpropamid is a novel rice blast controlling agent having two unique modes of action. In addition, carpropamid has unique physico-chemical properties which led us to develop it for nursery box application. This one shot application for nursery box prior to transplanting controls leaf as well as panicle blast.

Acknowledgements

We would like to express our sincere thanks to a number of colleagues at Bayer AG, Landwirtschaftszentrum Monheim, Germany and Nihon Bayer Agrochem K.K., Yuki Research Centre, Japan, for supporting this research.

References

1. Hunter, S. E.; Wye, A.; Boyce, C.; Armitage, B. P.; Haken, P.; Wagner, W. M. (Shell Int. Res.): German Pat. Appl. DE 2219710 (1972).
2. Cartwright, D.; Langcake, P.; Pryce, R. J.; Leworthy D. P.; Ride, J. P.: *Nature* **1977,** *267*, 511-513.
3. Kurahashi, Y.; Shiokawa, K.; Kagabu, S.; Sakawa, K.; Moriya, K. (Nihon Tokushu Noyaku Seizo K. K.): Jpn. Kokai Tokkyo Koho JP61015867, 1986.
4. Kagabu, S.; Kurahashi,Y.: *J. Pesticide Sci.* **1998,** *23*, 145-147.
5. Kraatz, U.; Hänßler, G. (Bayer AG): Eur. Pat. Appl. EP341475, 1990.
6. Kraatz, U.; Littmann, M.: *Pflanzenschutz-Nachr. Bayer (Ger. Ed.)* **1998,** *51*, 203-208.
7. Born, L. (Bayer AG), *Unpablished data*, 1987.
8. Merz, W.; Littmann, M.; Kraatz, U.; Mannheims,Ch. (Bayer AG): PCT Int. Appl. WO 9749665, 1997.
9. Kurahashi, Y.; Hattori, T.; Kagabu, S.; Ponzen, R.: *Pesticide Sci.* **1996,** *47*, 199-202.
10. Kurahashi, Y.; Araki, Y.; Kinbara, T.; Ponzen, R.; Yamaguchi, I.: *J. Pesticide Sci.* **1998,** *23*, 22- 28.
11. Nakasako, M.; Motoyama, T.; Kurahashi, Y.; Yamaguchi, I.: *Biochemistry* **1998,** *37*, 9931- 9939.
12. Araki, Y.; Kurahashi, Y.: *J. Pesticide Sci.* **1999,** *24*, 369-374.

Natural Products

The mind cannot fail to wonder at the beauty of the world's fauna and flora that verge on the miraculous in their diversity. It is the result of eons where species have appeared and disappeared. Each has been driven by a series of complex biological and physical events all working in concert to ensure successful candidates. Within each microcosm, be it animal or plant, are sequences of natural products that play vital roles. Some are regarded as secondary metabolites because we have not yet assigned any important functional role *in vivo*, but they are of primary importance in the fields of agriculture and medicine.

Discoveries in the field of natural products may be regarded as both macro and micro, the former being the discovery of individual biologically active molecules which stand out as gigantic discoveries, and the latter, their sets of congeners. An example illustrating this point is the discovery of the brassinosteroids starting, first, with brassinolide from canola pollen (*Brassica napus*), an important plant growth regulator, followed by the discovery of congeners in plant organs and then the elucidation of segments of the biosynthetic pathways.

We have not yet plumbed the depths of terrestrial natural product structures. The discovery of new life forms under Lake Vostok, the isolation of thermophiles and cryophiles, and the remarkably fast mutations that microorganisms undergo to produce new strains and, therefore, new natural products, means that plenty remains to be discovered and explained. Mother Nature, the great genetic engineer, has much to tell us about natural products in the future. The journey has only just begun.

Hiroshi Abe
Dept. of Applied Biological Science
Tokyo University of Agric. & Tech.
Fuchu
Tokyo 183-8509 JAPAN

Horace G. Cutler
Southern School of Pharmacy
Mercer University
3001 Mercer University Drive
Atlanta, GA 30341-4155

Chapter 5

Natural Products from New Zealand Plants

Nigel B. Perry

Plant Extracts Research Unit, New Zealand Institute for Crop and Food
Research Limited, Department of Chemistry,
University of Otago, Box 56, Dunedin, New Zealand

The various steps in discovering new bioactive natural products
are illustrated with examples from New Zealand's unique native
plants. This work is not restricted to the usual vascular plants,
but extends to liverworts and lichens. Fungicidal, herbicidal,
insecticidal and nematicidal compounds are described. The
bioactivity-directed isolation and structure elucidation processes
are detailed for two examples: a chlorinated fungicidal compound from a lichen; and a new insecticidal epoxy-ether from
a liverwort.

Natural products are an important source of new crop protection agents (*1-3*).
Many are of microbial origin, but a wide range of plant secondary metabolites also
have pest control properties (*4*). The usual steps in discovering new biologically
active (bioactive) natural products from plants are:

- Collecting
- Screening
- Bioactivity-directed isolation
- Structure elucidation

In this chapter, these steps are illustrated with examples from New Zealand's unique native plants. This work is not restricted to the usual vascular plants, but extends to liverworts and lichens, which are much less well studied for bioactive compounds. Natural products with potential for development as fungicidal, herbicidal, insecticidal and nematicidal crop protection agents are covered. The bioactivity-directed isolation and structure elucidation process is described in detail for a chlorinated lichen metabolite with fungicidal activity, and for a new insecticidal sesquiterpene epoxy-ether from a liverwort.

Collecting: Biodiversity and Chemical Diversity

The ideal of any bioactivity discovery program is to find new structures with new mechanisms of action. The chances of finding new structures are maximised by collecting organisms as different as possible from those that have been studied previously.

New Zealand is remote from the continents and was the last major land mass to be settled by humans when Maori arrived from Polynesia about 1000 years ago (5). The New Zealand archipelago extends from the subtropical Kermadec Islands (32° S) to the subantarctic Auckland and Campbell Islands (latitude 52° S). Plants have adapted to a range of environments including extensive subalpine zones, geothermal areas and serpentine soils (6).

This isolation and geographical diversity has led to considerable biodiversity. There are approximately 2300 native vascular plant species, of which 85% are endemic (Table I) (7).

Table I. New Zealand's Plant Biodiversity

Taxonomic Group	Species	Endemic
Vascular plants	2,300	81%
Ferns and allies	200	46%
Gymnosperms (conifers)	20	100%
Angiosperms	2,100	84%
Bryophytes (liverworts and mosses)	1,100	20-40%
Lichens	1,800	?

SOURCE: Based on (7).

At higher taxonomic levels about 35 New Zealand plant genera are endemic, and several more have only one or two outlying species in Australia or South America

(6). These unique plants have yielded many unusual natural products, which have been summarised by Cambie (8).

The generally high rainfall and unpolluted environment makes New Zealand a rich source of bryophytes and lichens (Table I). Bryophytes (9), especially liverworts, and lichens (10) are rich in secondary metabolites, presumably to protect these simple, soft organisms from herbivores and pathogens. However, the bioactivity of these metabolites is largely unexplored. Farnsworth (quoted by Verpoorte (11)) has stated that over 13,000 species of vascular plants have been studied for biological activity, whereas only 39 bryophyte and 118 lichen species have been similarly treated. Therefore we have worked with specialists to include these groups in our collections.

When planning collections, it should be borne in mind that a plant's secondary metabolism may vary both within the plant and over time. Temporal variation is most likely to be observed during rapid growth, especially of herbs (12). Different parts of vascular plants contain different amounts of secondary metabolites, but making separate collections of different plant parts increases screening costs. At Crop & Food Research we screen combined plant part samples, e.g. leaves, bark and twigs of trees, entire herbs or grasses, then extract separate parts if bioactivity is detected.

Fruits and seeds are often particularly rich in secondary metabolites. For example, we have found that the level of the insecticidal compound polygodial 1 (13) in fruits of *Pseudowintera colorata* (taxonomic information for most of the species mentioned in this chapter is given in Table II) is five times the level in foliage (Larsen and Lorimer, unpublished).

Another source of infraspecific variation is the presence of chemical races. These "chemotypes" usually involve switches within the same pathway of secondary metabolism. For example, polygodial 1 is the main sesquiterpene dialdehyde in some individuals and populations of *P. colorata*, but other individuals contain almost equal amounts of 1 and 2 (14). Both of these compounds have similar insecticidal activities (13) so this variation may not have adaptive significance for protection against herbivores.

However, it is also possible to have presence/absence of a class of bioactive compounds. For example, the herbicidal triketone leptospermone 3 (see below) is a major volatile component of foliage of *Leptospermum scoparium* from the East Cape of the North Island of New Zealand, but is scarcely detectable in the same species growing elsewhere in New Zealand and in Australia (*15*). This source of variation is ignored in most screening programs, but we regularly collect and screen the "same" species from different areas. Representative voucher specimens of each plant collection must always be kept for future reference.

Screening

Once a collection of plants is underway, the next step is to decide upon the biological assays to be used for screening, and the procedures for preparing plant samples for these assays. Most modern programs aim to test the full range of chemical diversity present in their collections, not restricting themselves to a particular class of compounds such as alkaloids. Different ways of achieving this are illustrated by the approaches taken in New Zealand screening programs:

- Singh *et al.* included milled leaf powders in insect diets to test for insecticidal activity of conifer foliage (*16*). This direct use of plant material has ecological relevance but preparation of solvent extracts is usual, since these can be tested in a wide range of different assays.
- Calder *et al.* used Soxhlet extracts in screening for fungicidal and antibacterial activities (*17*). This convenient continuous extraction procedure involves prolonged boiling of the extract and can lead to artefact formation. For example, an antifungal compound from a lichen (*18*) may have been produced during extraction with boiling methanol (*19*).
- The Crop & Food Research program uses room temperature extraction with 19:1 ethanol:water (rectified spirits) (*20*). This has the advantage of extracting compounds with a wide range of polarities in one step, from water-soluble glycosides (*21*) to volatile, non-polar ethers such as hodgsonox (see below). The disadvantage is that polyphenolics are also extracted and these give non-specific "nuisance" activity in some assays, especially enzyme-based ones. These can be removed simply by filtering through polyvinylpoly-pyrrolidone (*22*).

The choice of biological assays is vital to the success of a screening program. Researchers looking for crop protection agents have the luxury of being able to use *in vivo* assays at an early stage, whereas this is neither practical nor ethical for human conditions. The use of whole organisms allows for the discovery of new mechanisms of action, which are specifically excluded by the mode of action assays favored by pharmaceutical companies. On the other hand, whole organism assays are harder to miniaturize and automate for high throughput screening.

Because of the importance of agriculture and horticulture to New Zealand's economy, several groups have used local pests when screening the native flora for new crop protection agents. These assays also provide general indications of useful biological activity, as shown in the examples below.

Fungicides

There are many examples of higher plant secondary metabolites with antifungal properties, but most of these do not show effects *in vivo* (*23*).

The only published screening results for New Zealand plant extracts against plant pathogens are by Calder *et al.* (*17*). Three lichen extracts showed fungicidal activity against *Botrytis cinerea*.

Crop & Food Research, working with Dow AgroSciences, has also found fungicidal activity in New Zealand lichen extracts. Extracts of *Pseudocyphellaria granulata* and *P. faveolata* were active against *Phytophthora* and *Pythium* species *in vitro*, and were also active against *Plasmopara viticola*, *in vivo*. An extract of *P. granulata* was therefore selected for bioactivity-directed isolation work. Since we use ethanol to extract compounds with a wide range of polarities, we also need a primary fractionation method that can cope with this range of polarities. This is provided by reversed-phase (RP) flash chromatography (*24*). The extract was pre-coated onto RP material, then loaded onto a RP column, pre-conditioned with H_2O. The column was developed in steps from H_2O through MeCN to $CHCl_3$. We use acetonitrile rather than methanol because we have had problems with artefact formation by nucleophilic attack of methanol (Perry *et al.*, unpublished). All the fractions were screened for activity against *Phytophthora cinnamomi*, *in vitro* (Figure 1).

In this case much of the mass was recovered in low polarity fractions, probably containing the triterpenes known in this species (*25*). However, the fungicidal bioactivity was concentrated in fractions eluted with 1:3 and 1:9 H_2O:MeCN, and MeCN (Figure 1). Even more active was a pure compound that crystallized from the most active RP fraction (Perry & Brennan, unpublished). Bioactivity-directed isolations are not normally this simple!

The mass spectrum of this fungicidal compound showed the classic 3:1 isotope pattern of $^{35}Cl/^{37}Cl$ in a mono-chlorinated compound. The fragmentation matched that published for physciosporin 4 (*26*), a known metabolite of both P. granulata and P. faveolata (*25*). Chlorinated depsidones such as 4 are quite common amongst lichen metabolites (*27*), but no biological activity has been

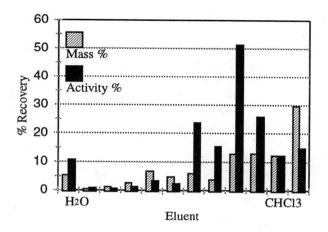

Figure 1. RP fractions from Pseudocyphellaria granulata: mass and antifungal activity against Phytophthora cinnamomi

reported for physciosporin. Further testing of **4** *in vivo* against *Plasmopara viticola* showed 49% disease control at 400 ppm, which was weak relative to commercial standards. However, there is clearly potential for discovery of new bioactivities amongst the many known lichen metabolites (*10*).

Herbicides

Herbicides make up almost half of the world agrochemical sales (*28*), so herbicidal compounds are important targets for screening programs. Plants have not been a good source of phytotoxins with potential for herbicide development, perhaps because of autotoxicity (*29*).

Crop & Food Research has found activity for New Zealand plant extracts in an assay on lettuce seed germination and seedling growth (*30*). Most of the extracts, from the foliage of endemic conifer species, significantly inhibited both germination and growth.

The triketone leptospermone **3** is a New Zealand plant metabolite that has been shown to have herbicidal activity. It is present at high levels in a regional chemotype of the shrub *Leptospermum scoparium* (*15*). In a study of herbicidal compounds from another plant in the Myrtaceae, Zeneca Ag Products found that **3** produced bleaching (*31*). Studies on related synthetic compounds have pointed to inhibition of *p*-hydroxyphenylpyruvate dioxygenase as the mode of action (*31*). A plant growth regulator G$_3$ **5** from an Australian myrtaceous species, *Eucalyptus grandis,* is structurally related to leptospermone **3** (*32*).

Insecticides

Plant secondary metabolites have played a major role in the development of commercial insecticides. The most important example is the development of the synthetic pyrethroids from the natural product pyrethrin (*23*).

Several research groups have screened New Zealand plants for insecticidal activity:

- Russell and co-workers found activity in most species of conifers (*16*) and in many ferns (*33*).
- Industrial Research Ltd. worked with Zeneca Agrochemicals on a range of plants (Bloor, unpublished).
- Crop & Food Research and AgResearch have screened a library of extracts against blowfly larvae (*34*).

Chemical diversity from gymnosperms

All of the 20 New Zealand gymnosperm species are endemic (Table I) and three of the genera are also endemic (*35*). These trees and shrubs have yielded insecticidal compounds from several different pathways of secondary metabolism (see Table II for source species):

- Triterpene-derived β-ecdysone **6** and related compounds (*16,36*).
- A norditerpene lactone, nagilactone C **7** (*37*).
- A lignan, β-peltatin-A methyl ether **8** (*38*).
- An alkaloid, dyshomoerythrine **9** (*39*).

Some of these compounds showed different modes of action in a blowfly larval development assay in which both survival and development were scored (*34*). Phytoecdysone **6** showed insect growth regulatory activity, with mortality of larvae increasing to a peak, then declining at higher concentrations. Nagilactone C **7** had growth inhibitory effects at sublethal concentrations. Dyshomoerythrine **9** showed a simple sigmoidal dose-response curve for mortality.

An epoxy-ether from a liverwort

Liverworts are rich sources of new secondary metabolites (*9*), and we have included New Zealand liverworts in our screening program. Extracts from *Lepidolaena hodgsoniae*, first collected from the subantarctic Campbell Island (*20*), showed activity against blowfly larvae. Bioactivity-directed isolation, using RP flash chromatography then silica, led to a single compound as the insecticidal component.

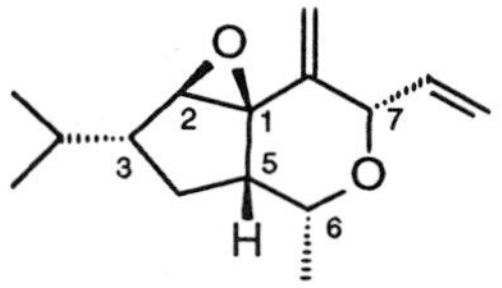

MS and NMR spectra gave the molecular formula $C_{15}H_{22}O_2$. The IR spectrum showed neither hydroxyl nor carbonyl groups and signals in the ^{13}C NMR spectrum suggested that the two oxygen atoms were incorporated in ether linkages. The compound was sufficiently volatile and stable to allow GC analysis. ^{13}C NMR data showed two C=C double bonds, so the compound was tricyclic. The structure was solved using 2D NMR experiments (COSY, HMQC and HMBC) to show ^1H-^1H and ^1H-^{13}C correlations. This process was complicated by the overlap of several ^1H signals, even at high field strengths and in a variety of solvents. Eventually a structure with a unique combination of bridgehead epoxide, 6-membered ether ring, exocylic methylene and vinyl groups was proposed for this new compound, nicknamed hodgsonox 10.

This structure contains six chiral centers. The relative stereochemistry was deduced by a combination of NMR nuclear Overhauser effect (NOE) experiments, which show which protons are close in space, and molecular modeling of the flexible rings and rotatable substituents. The NOE data indicated that H-3, H-5, H-6 and H-7 all lay on one face of the molecule (Figure 2). The stereochemistry of the epoxide was not settled by NOE experiments.

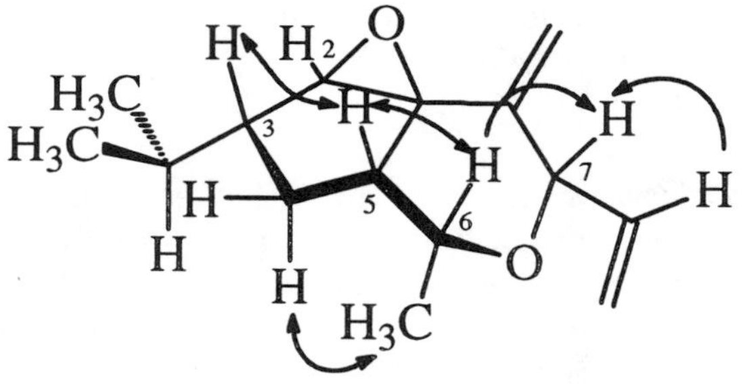

Figure 2. Predicted most stable conformation of hodgsonox 10, showing key NOE interactions.

Conformational searching and molecular modeling (40) was carried out on the two possible epoxides. The most stable conformation for the β-epoxide (Figure 2) gave a smaller predicted $J_{2,3}$ coupling than did the α-isomer, in line with the observation of a sharp singlet for H-2 in the ^1H NMR spectrum of 10. The observed H-5/H-6 coupling constant ($J_{5,6}$=4 Hz) was also closer to that predicted for the β-epoxide ($J_{5,6}$=3.7) than for the α-epoxide ($J_{5,6}$=5.3). These results led to the proposed relative stereochemistry of hodgsonox **10** (Lorimer et al., unpublished).

Hodgsonox **10** represents a new class of sesquiterpene featuring major rearrangements in its biosynthesis. The combination of a monosubstituted double bond and a 1,1-disubstituted double bond in a doubly allylic ether arrangement is novel, and may be related to its insecticidal activity. Hodgsonox **10** had an LC_{50} of 270 ppm against larvae of the blowfly *Lucilia cuprina*. This is less active than existing agents (e.g. diazinon LC_{50} 1.6 ppm) but the unique structure and unknown mode of action warrant further study.

Nematicides

Nematicides represent only a small share of the world crop protection market (*28*), but nematodes are economically important parasites of sheep and of root vegetables in New Zealand.

Crop & Food Research and AgResearch are screening our library of extracts against nematodes. An extract of a liverwort, *Plagiochila stephensoniana*, inhibited the motility of larvae of the sheep nematode *Trichostrongylus colubriformis* (*22*). This activity was attributed to the bibenzyl **11**, which had an

OMe **11**   OMe **12**

IC_{50} of 130 ppm. Bibenzyls are unusual amongst vascular plants, but comprise one of the major classes of secondary metabolites from liverworts (*9*). Synthetic studies (*41*) yielded a related stilbene **12** with a significantly lower IC_{50} of 60 ppm (*22*).

We are now working on a nematode larval development assay suitable for automated reading and therefore high throughput screening (Pernthaner *et al.*, unpublished).

Summary

These examples show that natural products with crop protection potential can come from diverse plant sources and biosynthetic pathways (Table II).

The future is bright for natural product discovery, as the full range of the world's biodiversity will be explored through the combination of fast, miniaturized assays and rapid compound identification, using techniques such as LC-MS and LC-NMR (*45*). Natural products with new modes of action will provide leads for synthetic optimization, using the power of combinatorial chemistry. Another development avenue for new natural products is the use of extracts as crop protection agents. One current example is the widespread use of extracts from the neem tree, *Azadirachta indica*, which contains the complex insecticidal triterpene azadirachtin (*46*).

Table II. Chemical diversity: pathways and plant sources

Pathway and Class	Cpd	Activity[a]	Source species[b]	Group (Family)
Isoprenoid				
Sesquiterpene	1	I, N	*Pseudowintera colorata*	Angiosperm (Winteraceae)
Sesquiterpene	10	I	*Lepidolaena hodgsoniae*	Liverwort
Diterpene	7	I	*Podocarpus spp.*	Gymnosperm (Podocarpaceae)
Triterpene	6	I	*Lepidothamnus spp.*	Gymnosperm (Podocarpaceae)
Polyketide				
Triketone[c]	3	H	*Leptospermum scoparium*	Angiosperm (Myrtaceae)
Depsidone	4	F	*Pseudocyphellaria granulata*	Lichen
Lignan	8	I	*Libocedrus bidwillii*	Gymnosperm (Cupressaceae)
Bibenzyl	11	N	*Plagiochila stephensoniana*	Liverwort
Alkaloid	9	I	*Lagarostrobos colensoi*	Gymnosperm (Podocarpaceae)

[a] Fungicidal, Herbicidal, Insecticidal or Nematicidal.
[b] For full taxonomic information see: *25,42-44*.
[c] Pathway not yet confirmed.

Acknowledgments

I thank my colleagues from a variety of disciplines, especially for permission to include results not yet published : chemists G. Ainge, N. Brennan, E. Burgess, S. Hinkley, L. Larsen, S. Lorimer, J. van Klink and R. Weavers; botanists A. Evans, D. Galloway and R. Tangney; and biologists E. Chapin, P. Douch, P. Gerard, A. Heath, A. Pernthaner and C. Snipes. I also thank the Department of Conservation, Dunedin City Council and various landowners for permissions to collect; and S. Lorimer, R. Weavers and T. Williams for suggestions on this chapter. Crop & Food Research's bioactive natural products work is supported by the New Zealand Foundation for Research, Science and Technology.

References

1. *Biologically active natural products: agrochemicals*; Cutler, H. G. and Cutler, S. J., Ed.; CRC Press: FL, 1999.
2. *Bioregulators for Crop Protection and Pest Control*; Hedin, P. A., Ed.; Amer. Chem. Soc.: DC, 1994.
3. *Crop Protection Agents from Nature: Natural products and analogues*; Copping, L. G., Ed.; Royal Society of Chemistry: Cambridge, 1996.
4. Grainge, M.; Ahmed, S. *Handbook of Plants with Pest-Control Properties*; Wiley: NY, 1987.
5. Enting, B.; Molloy, L. *The ancient islands. New Zealand's natural environments*; Port Nicholson Press: Wellington, 1982.
6. Wardle, P. *Vegetation of New Zealand*; Cambridge University Press: Cambridge, 1991.
7. Taylor, R.; Smith, I. *The state of New Zealand's environment 1997*; GP Publications: Wellington, 1997.
8. Cambie, R. C. *J. Royal Soc. N. Z.* **1996**, *26*, 483-527.
9. Asakawa, Y. *Progr. Chem. Org. Nat. Prod.* **1995**, *65*, 1-525.
10. Huneck, S.; Yoshimura, I. *Identification of lichen substances*; Springer: Berlin, 1996.
11. Verpoorte, R. In *Bioassay methods in natural products research and drug development*; Bohlin, L. and Bruhn, J. G., Ed.; Kluwer Academic: Dordrecht, 1999; Vol. 43, pp 11-23.
12. Perry, N. B.; Anderson, R. E.; Brennan, N. J.; Douglas, M. H.; Heaney, A. J.; McGimpsey, J. A.; Smallfield, B. M. *J. Agric. Food Chem.* **1999**, *47*, 2048-2054.

60

13. Gerard, P. J.; Perry, N. B.; Ruf, L. D.; Foster, L. M. *Bull. Entomol. Res.* **1993**, *83*, 547-552.
14. Perry, N. B.; Foster, L. M.; Lorimer, S. D. *Phytochemistry* **1996**, *43*, 1201-1203.
15. Perry, N. B.; Brennan, N. J.; van Klink, J. W.; Harris, W.; Douglas, M. H.; McGimpsey, J. A.; Smallfield, B. M.; Anderson, R. E. *Phytochemistry* **1997**, *44*, 1485-1494.
16. Singh, P.; Fenemore, P. G.; Dugdale, J. S.; Russell, G. B. *Biochem. Syst. Ecol.* **1978**, *6*, 103-106.
17. Calder, V. L.; Cole, A. L. J.; Walker, J. R. L. *J. Royal Soc. N. Z.* **1986**, *16*, 169-181.
18. Hickey, B. J.; Lumsden, A. J.; Cole, A. L. J.; Walker, J. R. L. *N. Z. Nat. Sci.* **1990**, *17*, 49-53.
19. Hylands, P. J.; Ingolfsdottir, K. *Phytochemistry* **1985**, *24*, 127-129.
20. Lorimer, S. D.; Barns, G.; Evans, A. C.; Foster, L. M.; May, B. C. H.; Perry, N. B.; Tangney, R. S. *Phytomedicine* **1996**, *2*, 327-333.
21. Lorimer, S. D.; Mawson, S. D.; Perry, N. B.; Weavers, R. T. *Tetrahedron* **1995**, *51*, 7287-7300.
22. Lorimer, S. D.; Perry, N. B.; Foster, L. M.; Burgess, E. J.; Douch, P. G. C.; Hamilton, M. C.; Donaghy, M. J.; McGregor, R. A. *J. Agric. Food Chem.* **1996**, *44*, 2842-2845.
23. Benner, J. P. In *Crop Protection Agents from Nature: Natural products and analogues*; Copping, L. G., Ed.; Royal Society of Chemistry: Cambridge, 1996, pp 217-229.
24. Blunt, J. W.; Calder, V. L.; Fenwick, G. D.; Lake, R. J.; McCombs, J. D.; Munro, M. H. G.; Perry, N. B. *J. Nat. Prod.* **1987**, *50*, 290-292.
25. Galloway, D. J. *Bull. Br. Mus. nat. Hist. (Bot)* **1988**, *17*, 1-267.
26. Maass, W. S. G.; McInnes, A. G.; Smith, D. G.; Taylor, A. *Can. J. Chem.* **1977**, *55*, 2839-2844.
27. Gribble, G. W. *Progr. Chem. Org. Nat. Prod.* **1996**, *68*, 1-467.
28. Copping, L. G.; Hewitt, H. G. *Chemistry and mode of action of crop protection agents*; Royal Soc. Chemistry: Cambridge, 1998.
29. Duke, S. O.; Abbas, H. K.; Amagasa, T.; Tanaka, T. In *Crop Protection Agents from Nature: Natural products and analogues*; Copping, L. G., Ed.; Royal Society of Chemistry: Cambridge, 1996, pp 82-113.
30. Perry, N. B.; Foster, L. M.; Jameson, P. E. *N. Z. J. Bot.* **1995**, *33*, 565-568.
31. Lee, D. L.; Prisbylla, M. P.; Cromartie, T. H.; Dagarin, D. P.; Howard, S. W.; Provan, W. M.; Ellis, M. K.; Fraser, T.; Mutter, L. C. *Weed Science* **1997**, *45*, 601-609.
32. Baltas, M.; Benbakkar, M.; Gorrichon, L.; Zedde, C. *J. Chromatogr.* **1992**, *600*, 323-326.

33. Russell, G. B.; Fenemore, P. G. *N. Z. J. Sci.* **1971**, *14*, 31-35.
34. Gerard, P. J.; Ruf, L. D.; Lorimer, S. D.; Heath, A. G. C. *N. Z. J. Agric. Res.* **1997**, *40*, 261-267.
35. Salmon, J. T. *The native trees of New Zealand*; revised ed.; Reed Methuen: Auckland, 1986.
36. Russell, G. B.; Fenemore, P. G.; Horn, D. H. S.; Middleton, E. J. *Aust. J. Chem.* **1972**, *25*, 1935-1941.
37. Russell, G. B.; Fenemore, P. G.; Singh, P. *Aust. J. Biol. Sci.* **1972**, *25*, 1025-1029.
38. Russell, G. B.; Singh, P.; Fenemore, P. G. *Aust. J. Biol. Sci.* **1976**, *29*, 99-103.
39. Bloor, S. J.; Benner, J. P.; Irwin, D.; Boother, P. *Phytochemistry* **1996**, *41*, 801-802.
40. Hinkley, S. F.; Perry, N. B.; Weavers, R. T. *Phytochemistry* **1994**, *35*, 1489-1494.
41. Lorimer, S. D.; Perry, N. B.; Tangney, R. S. *J. Nat. Prod.* **1993**, *56*, 1444-1450.
42. Allan, H. H. *Flora of New Zealand. Indigenous Tracheophyta. Psilopsida, Lycopsida, Filicopsida, Gymnospermae, Dicotyledones*; DSIR: Wellington, 1961; Vol. I.
43. Connor, H. E.; Edgar, E. *N. Z. J. Bot.* **1987**, *25*, 115-170.
44. Allison, K. W.; Child, J. *The Liverworts of New Zealand*; University of Otago Press: Dunedin, 1975.
45. Wolfender, J.-L.; Rodriguez, S.; Hostettmann, K. *J. Chromatogr.* **1998**, *794*, 299-316.
46. Stone, R. *Science* **1992**, *255*, 1070-1071.

Chapter 6

Chemistry of Spore Germination Self-Inhibitors from the Plant Pathogenic Fungus *Colletotrichum fragariae*

Hisashi Miyagawa[1], Masafumi Inoue[2], Hisaki Yamanaka[1, 4], Tetsu Tsurushima[3], and Tamio Ueno[1]

[1]Division of Applied Life Sciences, Kyoto University, Kyoto 6068502, Japan
[2]Central Research Laboratories, Dainihon Jochugiku Company, Ltd., Toyonaka, Osaka 5610827, Japan
[3]Han-nan University, Matsubara, Osaka 5808502, Japan
[4]Current address: Alcoholic Beverage, Seasoning and Food Research Laboratories, Takara Shuzo Company, Ltd., Otsu, Shiga 5202193, Japan

Colletotrichum fragariae, the fungal pathogen of strawberry anthracnose, exhibits self-inhibition phenomenon in terms of spore germination. Germination is dependent on population density, and is inhibited at higher concentrations of spores in water. Inhibition is caused by germination-inhibiting substances which are leached from the spores. The active compounds were isolated from acetone extracts of a potato-sucrose agar culture, and their structures were determined to be (2*E*, 4*E*)-2,4-hexadienal, (*E*)- and (*Z*)-3-ethylidene-1,3-dihydroindole-3-one, (2*R*)-(3-indolyl)propionic acid, and a pair of isomers named colletofragarone A and B.

Filamentous fungi, a major class of plant pathogens, generally have a life cycle which includes contact of the spore with the host plant, followed by spore germination, invasion into the plant tissue, growth and propagation at the expense of host health, and the reproduction of spores. Of these steps, spore germination is a critical event. The elucidation of the mechanism of spore germination is therefore important, if new devices for crop protection are to be developed.

Although spore germination is regulated by a variety of physical factors, including temperature, moisture, and light, it has been proposed that chemical factors also play a role in the regulation of germination. One of the observations that suggests the presence of chemical factor(s) is referred to as the germination self-inhibition of spores. When the collected spores are suspended in water, germination is, in some cases, dependent on the concentration of the spores: while most of spores readily germinate at low concentrations, the germination is inhibited under more crowded conditions. It has been suggested that this self-inhibition is caused by some chemical factor(s), which are released from the spores themselves. These putative factors, referred to as germination self-inhibitors, have attracted the interests of numerous investigators, since an understanding of this phenomenon could lead to the discovery of new types of antifungal agents.

To date, self-inhibition of germination has been observed in over 60 fungal species. An important fungus group among these are the plant pathogenic fungi which cause anthracnose. Anthracnose fungi consist of various species of the genus *Colletotrichum*, a teleomorph of which is classified as the genus *Glomerella*. The earliest report of the germination self-inhibition for one of the anthracnose fungi appeared as early as 1910 (*1*). In this paper, we wish to describe the characterization of the chemical factors which are responsible for the germination self-inhibition of spores from an anthracnose fungus, *Colletotrichum fragariae* Brooks, a strawberry pathogen.

Population-dependence of Conidium Germination of C. fragariae (2)

According to mycological terminology, cells that are generally referred to as spores are termed conidia for the case of an imperfect (asexual) fungus, such as *Colletotrichum*. When the conidia of *C. fragariae* are collected from the culture and suspended in water, the relationship between the rate of germination and the concentration of the conidia is as shown in Figure 1. The germination of the conidia was inversely dependent on the concentration, *i.e.*, the rate of germination was high at low concentrations, but self-inhibition occurred at a concentration of 10^7 spores per mL, and the resulting rate decreased to nearly zero.

However, by washing the conidia repeatedly, the rate of germination was increased even at high density (Figure 2), which suggests that some inhibitory factors, or germination self-inhibitors, may have been present in the original conidial suspension. After the self-inhibitors were removed by washing, the inhibition was canceled, thus enabling the start of the germination process.

Characterization of Self-Inhibitors from *C. fragariae*

Based on the above described findings, a search for the inhibitory factors was initiated. This search, however, was complicated because the size of an anthracnose conidium is quite small and because the spores cannot be obtained in large amounts. Therefore, an attempt was made to identify the candidate compounds in the whole cultured media of the fungus, and the role of the isolated substances was then verified by determining whether or not those substances are present in a conidial suspension.

The subject fungus was cultured on potato sucrose agar plates for 2 weeks and the culture was then extracted with acetone. By solvent partitioning and chromatographic separation, compounds which inhibit germination were purified and those referred to as CF-1 to 6 were obtained. Further analysis showed that both CF-5 and 6 respectively consisted of two closely related isomers CF-5-1 and 2, and CF-6-1 and 2.

Compound CF-1 was a volatile compound. Since the action of self-inhibitors has been observed to be reversible and transient in general, the volatility of a compound is thought to be a favorable property, for the action of such compound can be lost simply by evaporation. The structure of CF-1 was determined to be (2*E*, 4*E*)-2,4-hexadienal (**1**) by spectroscopic analysis. The

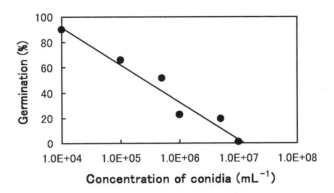

Figure 1. Relationship between germination rate and conidial concentrations of C. fragariae. (Reproduced from reference 2. Copyright 1996 Plenum Publishing Corporation)

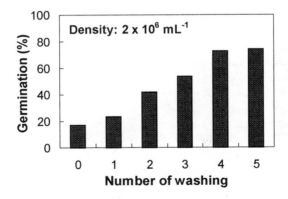

Figure 2. Effect of washing on the germination of C. fragariae conidia. (Reproduced from reference 2. Copyright 1996 Plenum Publishing Corporation)

presence of volatile self-inhibitors in fungal cultures has been described in several cases (*3, 4*). However, most have remained unidentified, and, as a result, this compound represents an important example of a volatile self-inhibitor.

The structures of CF-2 to CF-4 were also analyzed spectroscopically, and were determined to be a series of indole derivatives (**2 - 4**). They were identical to the self-inhibitors of the fungus of the same genus, *Colletotrichum gloeosporioides* f. sp. *jussiaea*, the structures of which had been previously determined by our group (*5*). This was the first isolation of **2** and **3** from a natural source, although they had already been synthesized (*6*). CF-4 has an *R* absolute configuration, and is the enantiomer of 2-(3-indolyl)propionic acid, which has been found earlier in cultures of *Claviceps purpurea* (*7*) and *Acremonium roseum* (*8*).

The self-inhibitor from *C. gloeosporioides* f. sp. *jussiaea*, gloeosporone (**5**) has also been reported (*9*), but it was not possible to verify the production of this compound in our study using the same strain. Moreover, the stereoselectively synthesized gloeosporone was not active as germination inhibitor (*5, 10*). Accordingly, the issue of whether gloeosporone is a true self-inhibitor is doubtful at present, although a considerable amount of effort concerning this compound has been made.

Compounds CF-5 and CF-6 respectively appeared as single peaks on normal phase HPLC using a silica gel column. However, when they were analyzed by reversed phase HPLC, each peak was separated into two peaks, and the compounds responsible for these peaks were designated, respectively, as CF-5-1 and 5-2, and CF-6-1 and 6-2. LC-MS analyses indicated that these compounds all have the same molecular weight, and produce M+H$^+$ ions at *m/z* 387. Since the MS fragmentation patterns were also similar to one another, it was likely that they were a series of isomers, or structurally-related derivatives. Among these, two compounds from the CF-6 fraction were so unstable that they decomposed into complicated product mixtures during purification by HPLC. CF-5-1 was also unstable but it did not undergo decomposition. Rather, it underwent isomerization to give CF-5-2. Of these compounds, CF-5-2 was the most stable, and was purified by reversed phase HPLC for spectroscopic characterization.

HR-MS indicated that the molecular formula of CF-5-2 was $C_{22}H_{26}O_6$. Other spectral data including HH- and CH-COSY indicated the presence of the following partial structures: a conjugated triene moiety; a chain structure containing 11 carbon atoms with two hydroxyl groups; a ketone moiety; an ester moiety and an allyl ether moiety. These partial structures were successfully combined with each other by long range CH coupling correlation data, which was collected by HMBC experiments, and the structure was proposed as **6**. The structure is unique, in that it contains a fused three ring-system including a ten-membered ring lactone. The entire geometrical structure of the olefinic side chain was determined to be *E*, based on proton coupling constants. This compound has not yet been described in the literature and, thus, is named colletofragarone A2 (*11*).

CF-5-1, which was thought to be an isomer of colletofragarone A2, gave a similar ^{1}H-NMR spectrum except in the olefinic proton region. Based on an analysis of the coupling constants of the olefinic protons, the structure was determined to be a geometrical isomer with the configurations of *E*, *Z*, *E* in the conjugated triene system (**7**). This compound is named colletofragarone A1 (*11*).

Another pair of compounds, CF-6-1 and 2, was very unstable, and therefore. the structures could not be determined. However, since LC-MS analysis indicated that their molecular formulae were the same as those of the colletofragarones, for which a large number of isomers are possible, the CF-6 compounds are very likely to be structurally related.

6 (CF-5-2)   7 (CF-5-1)

Biosynthesis of colletofragarone A2

The intriguing structure of the colletofragarones prompted us to study their biosynthesis. When the fungus was cultured with sodium acetate doubly-labeled with ^{13}C, the resulting colletofragarone A2 gave a ^{13}C-NMR spectrum, in which all of the carbon signals appeared as triplets. This was due to the overlapping of a singlet signal with a doublet signal, which resulted from the carbon-carbon coupling of the incorporated labeled acetate unit. This clearly demonstrates that all carbons of colletofragarone were derived from acetate. The arrangement of acetate units could be determined by reading the coupling constants of the incorporated carbon signals. Moreover, another feeding experiment using [1-^{13}C] acetic acid, followed by NMR analysis of the produced colletofragarone showed the increase of signal intensity for 11 of the 22 carbons which make up colletofragarone A2. Based on these results, the direction of each of the acetate units involved in the structure of colletofragarone could be determined as shown in Figure 3.

This labeling pattern indicates that the skeleton of colletofragarone is not constructed from a single polyketide chain, but, rather, is formed by the condensation of two precursor chains. Two possibilities for explaining this condensation (Figure 3) are: a pentaketide precursor combines with a hexaketide; a tetraketide combines with a heptaketide. However, the issue of which is the real pathway is unknown at present.

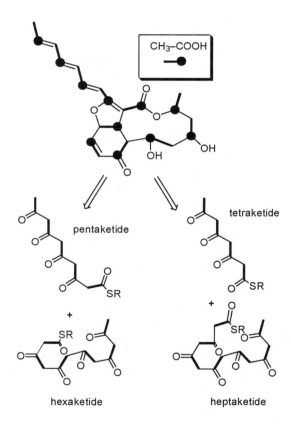

Figure 3. Incorporation of acetic acid in colletofragarone A2 and its possible precursors

Biological Activity of the Isolated Compounds

The germination inhibitory activity of compounds CF-1 to 6 against conidia of *C. fragariae* are shown in Table 1 in terms of the effective dose for a 50% inhibition, or ED_{50}. The value was 5 µg/ml for CF-2 and 3, while, for CF-1, 4, 5 and 6, they were somewhat larger, ranging from 20 to 35 µg/ml.

Table 1. Germination inhibitory activity of the isolated compounds

Compound	CF-1	CF-2	CF-3	CF-4	CF-5[1]	CF-6[1]
ED_{50} (µg/ml)	35	3	5	25	20	15

1 A mixture of isomers. The structure of CF-6 is unknown.

Among these inhibitors, the germination inhibitory activity of compounds CF-2 and 3 against the conidia of other fungal species was assayed (*5*). Interestingly, a slight level of selectivity was observed between the fungi that produce these compounds and those which do not. The conidial germination of *C. fragariae* and *C. gloeosporioides* was totally inhibited by CF-2 or 3 at 10 ppm, while concentrations of above 30 ppm were required to inhibit the conidial germination of other *Colletotrichum* species and *Fusarium oxysporum* f. sp. *cucumerinum*. However, a more strict selectivity has been observed for the germination self-inhibitors of rust fungi, another important fungal group, which has been extensively investigated in this respect. (*12, 13*)

Indole compounds CF-2 and 3 were not auxin-active, whereas CF-4 showed an activity comparable to that of indoleacetic acid, as evidenced by the *Avena* test. Although the involvement of auxins in the regulation of fungal growth has been a subject of considerable debate for many years, no definite conclusion has been drawn yet.

Analysis of Conidial Suspension Solution (*2*)

Conidia were collected from a 7 day old culture of *C. fragariae*, and dispersed in water at a concentration of 10^7 mL^{-1}. HPLC analysis showed the presence of CF-2, 3 and a pair of CF-5 isomers in this conidial suspension, thus demonstrating that these compounds are likely to be associated with conidial germination of this fungus. In addition, two further unknown peaks were detected, which were not identical with CF-6. Interestingly, both gave molecular ions of the same mass as those of CF-5 by LC-MS analysis and, therefore, appear to be the structurally related compounds. It is possible that these CF-5 related compounds were derived from a very unstable common precursor, which decomposes by several pathways depending on the conditions.

Possible Role of Self-inhibitors in Germination

The presence of self-inhibitory compounds in a conidial suspension suggests that they are responsible for the population-dependent germination described above. The observations may be explained as follows. The inhibitors are originally deposited in spores, or in the matrix surrounding the spores. When the spores are brought into contact with water, the inhibitors leach out. For a high concentration of spores, the concentration of inhibitor reaches levels sufficient to inhibit germination, while at lower concentrations of spores, their effect is negligible.

If this is the case, what is the physiological significance of these compounds for the fungus itself? From an ecological point of view, crowded conditions of the spores represent a competitive and unfavorable situation. Under these conditions, the inhibitors conveniently retard germination until conditions become less competitive for fungal growth. Such a function may also be advantageous when a spore find itself in a situation where only a limited amount of water is available. A more likely role for the function of a self-inhibitor, however, is to prevent germination, while the spores are still in the colony and do not need to germinate. In other words, the germination self-inhibitor protects the spores from untimely germination (14). Thus, the mode of action analysis of self-inhibitors may disclose the mechanism of release from dormancy to germination of fungal spores.

Concluding Remarks

A germination self-inhibitor has also been found in another anthracnose fungus, *C. graminicola* (3). The structure of the inhibitor, mycosporine-alanine, was determined to be **8** , which is completely different from those identified in our study. This suggests that self-inhibitors of anthracnose fungi are rather species dependent. If so, more structurally interesting compounds might be obtained by continuing the work with other anthracnose fungi.

8

The self-inhibitors hitherto identified are relatively unstable compounds in general. However, if their stability could be improved by chemical modification without losing inhibitory potency, and their mode of action examined in more detail using these improved inhibitors, we will be able to clarify yet-unknown aspects of the regulation mechanism of fungal life cycles. Such information should be important in exploring possible weaknesses of pathogens that could be exploited towards developing a new disease control strategy.

Acknowledgments

The strain of *Colletotrichum fragariae* used in this study was provided by Dr. G. Templeton of the University of Arkansas.

References

1. Edgerton, C. W. *La. Agric. Exp. Stn. Bull.* **1910**, 119, 1.
2. Inoue, M.; Mori, N.; Yamanaka, H.; Tsurushima, T.; Miyagawa, H.; Ueno, T. *J. Chem. Ecol.* **1996**, 22, 2111.
3. Leite, B.; Nicholson, R. J. *Exp. Mycol.* **1992**, 16, 76.
4. Leite, B.; Nicholson, R. J. *Mycologia* **1993**, 85, 945.
5. Tsurushima, T.; Ueno, T.; Fukami, H.; Irie H.; Inoue, M. *Mol. Plant-Microbe Interactions* **1995**, 8, 652.
6. Tacconi, G.; Maggi, L. D.; Rihgetti, P.; Desmoni, G. *J. Chem. Soc. Perkin II* **1975**, 150.
7. Yamano, T. *Nippon Nougeikagaku Kaishi* **1961**, 35, 1284 (*in Japanese*).
8. Yoshida, N.; Sassa, T. *Agric. Biol. Chem.* **1990**, 54, 2681.
9. Meyer, W. L.; Lax, A. R.; Templeton, G. E.; Brannone, M. J. *Tetrahedron Lett.* **1983**, 24, 5059.
10. Matsushita, M.; Yoshida, M.; Zhang, Y.; Miyashita, M.; Irie, H.; Ueno, T.; Tsurushima, T. *Chem. Pharm. Bull.* **1992**, 40, 524.
11. Inoue, M.; Takenaka, H.; Tsurushima, T.; Miyagawa, H.; Ueno, T. *Tetrahedron Lett.* **1996**, 37, 5731.
12. Macko, V; Staples, R. C.; Allen, P. J.; Renwick, J. A. A. *Science* **1971**, 173, 835
13. Tsurushima, T.; Mayama, S.; Tani, T.; Ueno, T; Fukami, H. *Ann. Phytopathol. Soc. Jpn.* **1984**, 50, 582.
14. Gottlieb, D. *Phytopathology* **1973**, 63, 1326.

Chapter 7

Bioherbicides: Phytotoxic Natural Products

Robert E. Hoagland

Southern Weed Science Research Unit, Agricultural Research Service,
U.S. Department of Agriculture, P.O. Box 350, Stoneville, MS 38776

Some plant and microbial secondary compounds possess a wide diversity of biological activity, including phytotoxicity. There is a broad range of potency and selective action of such natural phytotoxic products on plants.   Some natural phytotoxins are sufficiently efficacious to be used directly, while others may offer new chemical templates, useful in the design of new chemical classes of herbicides that act at new molecular sites. Various microbes produce phytotoxic compounds, however, phytopathogens use such compounds to injure or kill plants during pathogenesis. Phytotoxins from plants and microbes (and the pathogenic microbes themselves) can be classified as bioherbicides when they are used in weed control strategies.   The decline in the number of new synthetic herbicides derived from traditional screening programs (especially those with new molecular modes of action), and the increasing spread of weed resistance to synthetic herbicides, provides impetus for natural products as important alternatives for herbicide development. This overview examines some plant and microbial compounds that have potential as herbicides, and comments on future perspectives of bioherbicides.

Natural products have been used for thousands of years in attempts to control various pests, but the first use of such agents against weeds is uncertain. Synthetic herbicides have played a major role in revolutionizing food and fiber production over the past six decades. New compounds for herbicide development have traditionally come from synthesis and screening, but novel chemical approaches over the past 20 years have examined natural products from plants and microbes as structural templates for the design of new herbicides. Bioherbicides can be defined as plant pathogens and phytotoxins derived from plants, pathogens, and other microorganisms that have potential for weed control. Increased interest in the control of weeds with bioherbicides, and progress has been summarized, e.g., (*1-11*). Some weeds have developed resistance to synthetic herbicides, and herbicide mobility and/or persistence have resulted in the accumulation of some of these compounds in run-off and ground water. Some traditional herbicides are continually being removed from the market due to economic, toxicological, or other factors. Thus there is a need for new herbicides with more desirable properties, i.e., novel molecular modes of action, improved selectivity, and lower persistence and non-target toxicity. Herbicidal products derived from microbes and plants, and the use of pathogens may provide alternatives for improved weed control and environmental stewardship. This overview examines some compounds and future perspectives of bioherbicides as weed control agents. The use of pathogens for weed control is not discussed here, but has been reviewed (*6, 10, 11*).

Bioherbicidal Compounds from Plants

Artemisinin (I). Annual wormwood (*Artemisia annua*) leaves and flowers contain artemisinin, an antimalarial endoperoxide sesquiterpene lactone (*12*). Artemisinin is also phytotoxic and inhibits seed germination and growth of various seedlings (*13*). Other sesquiterpene lactones from *A. annua* are plant growth regulators or allelochemicals (*14, 15, 16*). Arteether, a reduction product of artemisinin, was phytotoxic to plant species at 1 ppm, and lower concentrations promoted plant growth (*13*). Radish (*Raphanus sativus*) and annual wormwood were highly tolerant to artemisinin (*13*). Artemisinin was autotoxic to annual wormwood at 33 ppm (*15*). Artemisinin is localized entirely in the subcuticular space of leaf capitate glands (*17*).

The mode of action of artemisinin is not fully understood, however data suggest that it acts by a mechanism other than disruption of mitosis, alteration of amino acid biosynthesis, or inhibition of respiration (*15*). Although the antimalarial of artemisinin in animal cell cultures is associated with large reductions in putrescine levels, plants treated with this compound displayed only slight reductions in polyamine levels (*18*). This suggested that the primary biochemical activity in plants is not blockage of the polyamine biosynthetic pathway. The effects of artemisinin on the plasma membrane of *Lemna* spp. also does not explain its inhibition of seed germination and growth (*19*). Preliminary data suggest that artemisinin may inactivate porphyrins (*3*). The complex structure of artemisinin renders it unsuitable as a commercial herbicide, however the oxabicyclo- heterocylic portion of the artemisinin molecule is common to the herbicide, endothall (7-oxabicyclo[2.2.1]heptane-2,3-dicarboxylic acid), and to the natural products, cantharidin and cineole (see below).

Cineole (II). 1,8-Cineole is one of the most potent phytotoxins isolated from plants that was initially studied for phytotoxicity and herbicidal potential. Cineole, which has structural features common to the natural plant product canthardin and the herbicide endothall, is produced by many plant species and plays a role in allelopathic interactions in some plants (*20*). Although cineole has potent phytotoxicity, its high volatility which is undesirable. An analog of cineole was synthesized, i.e., cinmethylin (**III**), which received some product development, but was ultimately not marketed as a herbicide in the U.S.(*21*), but it was sold in Europe. The mode of action of this chemistry is unknown, but it is a growth inhibitor of some grasses and broadleaf weeds, inhibits respiration (*22*), and inhibits mitosis in meristematic tissues (*23*). Cinmethylin which has structural characteristics similar to artemisinin, also had no dramatic effect on polyamine biosynthesis (*18*). Recent evidence showed that growth inhibition induced by cineole and cinmethylin was reversed by exogenous asparagine (*24*). Also, cineole strongly inhibited asparagine synthetase, but cinmethylin lacked potent inhibition. Data suggest that this enzyme is the site of action of these compounds, and that cinmethylin is a proherbicide that requires metabolic activation to become herbicidal.

Nitropropionic acid (IV). Hiptagenic acid was probably the first nitro-compound isolated from plants, and was later identified as β-nitropropionic acid (NPA) (*25*). NPA is also produced by the fungi, *Aspergillus flavus, A. oryzae, A. wentii,* and *Penicillium atrovenetum* (*26*), and by a pathogen of *Zinnia* spp. (*27*). *Septoria cirsii*, a host-specific pathogen of Canada thistle (*Cirsium arvense* L.), causes leaf spot, is an effective weed biocontrol agent, and produces NPA. NPA inhibited germination and induced chlorosis and necrosis in this host, and several other plant species were sensitive (*28*).

NPA was also phytotoxic to other weed and crop species (*29*). NPA was an ineffective germination inhibitor, but foliar applications reduced shoot elongation in some species. Excised leaf segments of dandelion (*Taraxacum officinale* Weber), curly dock (*Rumex crispus* L.), and crimson clover (*Trifolium incarnatum* L.) exhibited more chlorosis/necrosis when exposed to NPA (10^{-3}M) in the light than in the dark. In greening tests, NPA (10^{-3} M) completely blocked chlorophyll production.

NPA irreversibly inhibits succinate dehydrogenase (*30*), but inactivation is not due to nucleophilic addition of the β-position of nitropropionate to the flavin acceptor. The actual inactivator is the oxidation product of nitropropionate (*31*)

Sorgoleone (V). Sorgoleone, a major constituent of sorghum (*Sorghum bicolor*) root exudates (*32*), is a potent phytotoxin in a variety of species. Sorgoleone and two synthetic herbicides {metribuzin [4-amino-6-(1,1-dimethyl)-3-(methylthio)-1,2,4-triazin-5(4*H*)-one] and diuron [*N*′-(3,4-dichlorophenyl)-*N,N*-dimethylurea]} inhibited competitive binding with atrazine [6-chloro-*N*-ethyl-*N*′-(1-methylethyl)-1,3,5-triazine-2,4-diamine] in thylakoids of susceptible species, but not in thylakoids of resistant species (*33*). The binding affinity of sorgoleone was intermediate to that of diuron and metribuzin (*34*). Soil-incorporated sorgoleone (>40 ppm) reduced the fresh weight of shoots, but not roots. Sorgoleone is a potent inhibitor of electron transfer between Q_a to Q_b at the reducing side of PSII (*35, 36*).

Other Plant-Derived Phytotoxins. The alkaloids, colchicine and vinblastine, and the terpenoid taxol (structures not shown) are also phytotoxic and disrupt mitosis (37). These compounds have not been utilized in agriculture to a high degree, probably because of their high pharmaceutical value. Some higher plants produce potent

I. Artemisinin   II. Cineole   III. Cinmethylin

IV. Nitropropionic acid

V. Sorgoleone

Selected examples of bioherbicidal compounds from plants.

phytotoxic compounds that are photodynamically activated, e.g., alpha-terthienyl, phenylheptatriyne, hypericin, and fagopyrin (structures not shown)(38). Athough the use of photodynamic compounds as herbicides is a logical strategy, many of these compounds are toxic to many non-target species upon irradiation.

Bioherbicidal Compounds from Microorganisms

Microorganisms have produced more compounds with bioherbicidal potential than plants (2, 3, 6). Within the various classes of microorganisms, it might expected that pathogens would be richer sources than non-pathogens, but this may not be the case. Many genera from various bacterial families cause disease in plants. For example, in the family Pseudomonadaceae (comprised of the genera Pseudomonas and Xanthomonas), over 150 plant pathovars have been characterized. In contrast, more than 400 Streptomyces species are described, but only a couple are pathogenic. Streptomyces are diverse filamentous Gram-positive bacteria, that have produced many unique chemicals (39-42). Streptomyces species have been the most prolific microbial sources for the production of phytotoxic compounds.

Phytotoxins from Non-pathogenic *Streptomyces*

Bialaphos (VI) and Phosphinothricin (VII). Of all the numerous phytotoxic metabolites produced by *Streptomyces* species, bialaphos and phosphinothricin [marketed as glufosinate, the synthesized ammonium salt of phosphinothricin] constitute the most successful. Two laboratories discovered a naturally-occurring compound identified as bialaphos in 1971. This tripeptide contained a unique amino acid, L-2-amino-4-[hydroxy(methyl)phosphinyl]butyric acid (called phosphinothricin or PPT) linked to two L-alanyl moieties and was isolated from *S. viridochromogenes* (43), and from *S. hygroscopicus* cultures (44). The L-isomer (L-PPT) is the natural form, and was the first reported naturally-occurring amino acid containing a phosphinic group. It was initially found to have antifungal (*Botrytis cinerea*) and antibacterial activity (43). PPT also strongly inhibited glutamine synthetase activity in *Escherichia coli* (43). Later L-PPT was shown to have herbicidal activity, and these results were patented (45). Synthesis of the DL-PPT ammonium salt (common name, glufosinate) resulted in the commercial herbicide. Bialaphos exhibits phytotoxicity when L-PPT is released via hydrolytic cleavage of the alanine moieties. Bialaphos has been patented as a herbicide (46), and is marketed in Japan (47). GS inhibition and ammonia accumulation are generally thought to be pivotal to glufosinate mode of action (48), but other molecular sites have been suggested. Bialaphos does not inhibit GS, but peptidases in plant tissues convert it to PPT (49). D-PPT also does not inhibit GS (50), is non-herbicidal (51), and is not degraded in transgenic plants (52).

The complex biosynthetic pathway of bialaphos and PPT consists of over a dozen steps (53). One step involves an acetyl CoA-dependent reaction that modifies either demethyl-PPT or PPT. The gene (*bar*) responsible for encoding this step confers resistance to bialaphos in the organism. The gene has been isolated and characterized, and found to encode acetyl transferase activity which converts PPT to an acetylated non-phytotoxic metabolite (54). To date, many vegetable and cereal crop species have been transformed with either the *bar* gene of *S. viridochromogenes* or the *pat* gene of *S. hygroscopicus* (55-58). Presently more than 20 crop plant species have been transformed for resistance to PPT and some plants are resistant to PPT at rates up to10 times the lowest normal field application rate (55).

The anti-fungal activity of bialaphos and glufosinate was recently assessed on three pathogens *in vitro* and *in vivo* on PPT-resistant transgenic creeping bentgrass *(Agrostis palustris)*, an important turf grass (59). Bialaphos simultaneously controlled weeds and fungal pathogens in this transgenic grass. Bialaphos has antibiotic activity against *Rhizoctonia solani* Kühn that causes rice *(Oryza sativa)* sheath blight (60), and *Magnaporthe grisea* (Herbert) Barr (61) that causes rice blast disease. Substantial suppression of sheath blight symptoms occurred when bialaphos was applied to transgenic plants infected with *R. solani* prior to herbicide treatment (62). Inoculated transgenic [(*bar*) gene] rice plants had reduced disease symptoms after bialaphos treatment for weed and disease control (63). Bialaphos and PPT are unique herbicides since they have both potent antibiotic and herbicidal properties. This dual strategy will no doubt be utilized more widely with the increasing availability of PPT-resistant crops.

PPT and Bialaphos Analogs. Two oligopeptides containing PPT accumulated in cultures of a bialaphos producer, *S. hygroscopicus* SF1293, when bialaphos was added (64). One oligopeptide was a new compound [a bialaphos dimer (phosphinothricyl-ala-ala-phosphinothricyl-ala-ala)], and the other was previously known (phosphinothricyl-ala-ala-phosphinothricin). Other herbicidal peptide analogs of bialaphos have been isolated: phosalacine (PPT-ala-leu) from *Kitasatosporia phosalscinea* KA-338 (65, 66) and trialaphos (PPT-ala-ala-ala) from *S. hygroscopicus* sp. KSA-1285 (67).

Other naturally-occurring phytotoxic GS inhibitors are structurally similar to PPT (68). L-Methionine sulfoximine (MSO) was initially synthesized (69), then found in tree bark of *Cnestis glabra* (70). MSO was the first reported inhibitor of GS (71, 72). The PPT analog, L-(N^5-phosphono)methionine-S-sulfoximine (PMSO), is a potent GS inhibitor and a metabolite of L-(N^5-phosphono)methionine-S-sulfoximinyl-L-ala-L-ala (PMSO-Ala-Ala) (73) isolated from a *Streptomyces* sp.(74). PMSO and MSO can form from peptidase action on PMSO-Ala-Ala or phosphatase action on PMSO, respectively (75, 76). MSO has been patented as a herbicide (77), but its inhibitory activity on GS is much less than that of PPT (78). In whole plant tests, PPT was generally 5- to 10-fold more phytotoxic than MSO (79). Although numerous GS inhibitors that are analogs of PPT have been synthesized, none are more herbicidal than PPT (80-83).

Anisomycin (VIII). Anisomycin (from *Streptomyces* sp.) exhibited selective root growth inhibition at 12.5 ppm, and shoot growth inhibition at levels >50 ppm in monocotyledons [rice, barnyardgrass *(Echinochloa crus-galli)*, crabgrass *(Digitaria adscendens)*], and dicotyledons [leucerne *(Medicago sativa)* and tomato *(Lycopersicon esculentum)*] (84). The analog, acetyl-anisomycin inhibited shoot and root growth in rice and barnyardgrass more strongly than anisomycin. Another analog, *N*-acetyl anisomycin, showed little or no phytotoxicity at 100 ppm. The total synthesis of this antibiotic has been achieved (85). Derivatives based on this structure led to diphenylmethane analogs and eventually to 4-methoxy-3,3'-dimethylbenzophenone, a selective herbicide (NK-049; Nippon Kayaku Co.) for rice in 1974 (86). Anisomycin thus became the first microbial product that led to the development of a synthetically-based commercial herbicide. Phytotoxicity of PPT, a second natural product from a microbial source, was not patented until 1977 (87).

Herbicidins (IX). Several antibiotic analogs of herbicidin with selective contact herbicidal activity are produced by *Streptomyces saganonensis* (88). At least four of these analogs exhibit similar phytotoxicity spectra. These compounds were generally more active against dicotyledons; and among monocotyledons, rice was highly resistant to herbicidal injury. Herbicidin A was more selective between rice and the other plants tested than herbicidin B. These compounds had no activity on a variety of bacteria when tested at 100 μg/ml. Although *Xanthomonas oryzae*, *X. campestris*, *X. citri*, and *X. pruni* were not susceptible to these herbicidins at 100 μg/ml *in vitro*, application of herbicidin A or B to rice leaves could provide some effectiveness against *X. oryzae* (bacterial leaf blight disease) when sprayed at 3 ppm. Complete protection was afforded by spraying either 100 or 30 ppm herbicidin A or B, respectively.

Herboxidiene (X). A novel polyketide, herboxidiene, with potent phytotoxic activity on several annual weed species was isolated from *S. chromofuscus* A7847 (89). The microbial metabolite also inhibited *Arabidopsis thaliana* seed germination. In post-emergence tests (35 g/A in pot assays), it provided excellent control (≥ 90%) of broadleaf annual weeds including rape (*Brassica napus*), wild buckwheat (*Polygonum convolvulus*), and morningglory, and at 7 g/A, inhibited (25%) rape, wild buckwheat, and hemp sesbania. When rape was planted with wheat (*Triticum aestivum*), herboxidiene (35 g/A) controlled rape without wheat injury. This compound may be useful for weed control in rice and soybean, which are also insensitive (90). The absolute molecular configuration of herboxidiene has been elucidated (91).

Hydantocidin (XI). Hydantocidin is a relatively new phytotoxin discovered in the culture broth of *S. hygroscopicus* SANK 63584 (92). The compound was crystallized and found more phytotoxic against perennial than annual plants. Stereochemical confirmation and reconfirmation of the structure, involved experiments that provided total synthesis of hydantocidin and its stereoisomers (93-95). Herbicidal activity of this compound has been reported (96), and it was nonselective against mono- and dicotyledonous perennial and annuals. Deoxy-derivatives of hydantocidin have been synthesized, but had no phytotoxicity (94, 95). In comparative tests of hydantocidin, glyphosate [*N*-(phosphonomethyl)glycine], and bialaphos on 17 weed species, hydantocidin activity on annual monocots and dicots was equal to glyphosate, and slightly greater than bialaphos (96). Similar effects were found for these three compounds on perennial weeds. Hydantocidin had no antimicrobial activity when tested against 35 species of bacteria, yeasts, and fungi, even at 1000 μg/ml, and has low toxicity to mammals and fish. Due to its low non-target toxicity and high systemic herbicidal activity, the compound may have very high potential for development.

Hydantocidin is metabolized *in planta* to hydantocidin-5'-phosphate, which is a potent inhibitor of adenylosuccinate synthase (*97-99*). Growth inhibition caused by exposure to hydantocidin can be reversed when treated plants are supplied with adenine, AMP, or adenylsuccinate, but IMP was ineffective (*3*). Hydantocidin has no *in vitro* effect (*99*), but its 5'-phospho derivative is a potent adenylosuccinyl synthetase inhibitor. Hydantocidin blocks conversion of IMP to AMP (*98, 99*). Thus, hydantocidin-5'-phosphate appears to be the actual phytotoxin, and hydantocidin is a proherbicide. Crystal structure analysis of adenylosuccinate synthase bound to hydantocidin-5'-phosphate may lead to the design of more potent inhibitors (*100*).

Pyridazocidin (XII). Pyridazocidin, a new compound from cultures of a *Streptomyces* sp. had significant postemergence phytotoxicity on several weeds [velvetleaf (*Abutilon theophrasti*), ivyleaf morningglory (*Ipomoea hederaceae*), barnyardgrass, and giant foxtail (*Setaria faberi*)], and giant foxtail was most sensitive (*101*). Pyridazocidin caused rapid oxygen consumption in isolated chloroplasts over a concentration range that also inhibited plant growth. Phytotoxicity of the molecule (XII) is lost when the R-group is changed from a hydroxyl (OH = pyridazocidin) to an amine (NH$_2$). Pyridazocidin's mode of action of is similar to the bipyridinium herbicides (diquat, paraquat); i.e., inhibition of PS I.

Other Phytotoxins Discovered from *Streptomyces*. Many other compounds possessing phytotoxicity have been identified from *Streptomyces* species (Table I). These chemicals (structures not shown) vary in structure from simple to complex and they have demonstrated bioherbicidal potential on various plant species using bioassay techniques such as whole plant, excised leaf, coleoptile, seed germination, and pigment analysis (*41*). Several other recently discovered phytotoxins (structures not shown) are briefly mentioned below. Homoalanosine (*S. galilaeus*) controls some weeds at 4 kg/ha without injury to rice (*111*). *S. albus* produced a phytotoxin, AT-265,with potent post-emergence activity at > 100 kg/ha (*112*). However, high mammalian toxicity precludes its development as a herbicide. Vulgamycin (from *Streptomyces* sp.) had post-emergence activity, with no injury to barley (*Hordeum vulgare*), cotton (*Gossipium hirstum*), or corn (*Zea mays* L.) (*113*). Hydranthomycin, an anthroquinone, is a newly discovered chemical from *Streptomyces* sp. K93-5305 (*114*) with phytotoxicity to sorghum and radish, and the alga, *Euglena gracilis*.

Phytotoxins from Non-*Streptomyces* Microorganisms.

Cornexistin (XIII). Cornexistin was extracted from the culture broth of the basidiomycete *Paecilomyces variotii* SANK 21086 found on deer dung (*115*). It is a nonadride compound (nine membered carboxylic ring compound that has one or more five-membered rings). It has excellent postemergence herbicidal activity against dicot and monocot weeds, and corn is tolerant. It is apparently metabolized *in planta* to a compound that inhibits aspartate amino transferase (*116*). Cornexistin has no effect on

Selected examples of bioherbicidal compounds from *Streptomyces*.

asparate amino transferase *in vitro*, but inhibition does occur during incubation with cell-free plant extracts. It appears to be a proherbicide, and a metabolite similar to gostatin (a known aspartate amino transferase inhibitor) is thought to be formed (*117*).

Cyperin (XIV). Cyperin has been isolated from fungal cultures of *Preussia fleischhakii* (*118*), *Ascochyta cypericola* (*119*) and *Phoma sorghina* (*120*); the latter two are pathogens of purple nutsedge (*Cyperus rotundus*) and pokeweed (*Phytolacca americana*), respectively. This diphenyl ether was highly phytotoxic to purple nutsedge (*1119*), and several other plant species (*120*). Other studies showed that cyperin inhibited purple nutsedge root growth grown on agar, but not in soil (*121*). It also inhibited growth of mouse-ear cress (*Arabidopsis thaliana*) and marsh bentgrass (*Agrostis palustris*). Cyperin disrupts membranes, causes growth inhibition (*121*), lowers chlorophyll content, and is a weak inhibitor of protoporphyringogen oxidase (*3*). Results suggest that the mode of action of cyperin differs from that of synthetic herbicides such as acifluorfen, which inhibits protoporphyrinogen oxidase.

Other Examples of Microbial Phytotoxins with Bioherbicidal Activity. Other selected examples of phytotoxins (structures not shown) from other microbial species are shown in Table II. Moniliformin was used as a herbicide template and many analogs were synthesized in attempts to find commercial herbicides (*4*). Compactin and mevinolin (6-alpha-methylcompactin) are potent inhibitors of the isoprenoid pathway in plants (*136*). They exhibit strong phytotoxicity and act by inhibiting 3-hydroxy-3-methylglutaryl CoA reductase, the key enzyme of the isoprenoid pathway (*137*).

Microorganisms also produce photodynamic phytotoxins such as cercosporin (*Cercospora kikuchii*) which causes purple speck disease in soybean (*138*) and dothistromin (*Dothistroma pini*) the causal agent of blight in *Pinus* (*139*). Many of these compounds have not been pursued as herbicides because of toxicity or nucleic acid damage to non-target organisms upon exposure to light.

Summary and Perspectives

Not unexpectedly, plants do not produce numerous potent phytotoxic chemicals. With regard to microorganisms, phytopathogens should be expected to be good sources of phytotoxins that could provide leads for bioherbicide development. Most of our knowledge of pathogen-plant interactions is centered on crop plants and their associated pathogens. Comparatively, there is little information on the biochemical interactions of pathogens that infect weeds, and in many cases the phytotoxins from weed pathogens have not been identified (*140*). Furthermore, many non-pathogenic fungi and bacteria, especially *Streptomyces* species, have produced an abundance of phytotoxic compounds. Thus, there appear to be many untapped sources of important new bioherbicidal chemistries. Some of the plant and microbial compounds examined here have been

Table I. Examples of Other Compounds from _Streptomyces_ with Bioherbicidal Activity.

Compound	Source	Reference
Altemicidin	_Streptomyces_ sp.	_(102)_
Azaserine	"	_(103)_
Cispentacin	"	_(102)_
Coaristeromycin	"	_(104)_
7-Deoxy-D-glucoheptose	"	_(105)_
2,5-Dihydrophenylalanine	"	(102)
Filipin	_S. filipenensis_	_(106)_
Gougerotin	_S._ sp.	_(107)_
Isoxazole-4-carboxylic acid	"	_(108)_
Methelene-b-alanine	"	_(102)_
Naramycin	_S. griseus_	_(109)_
Toyocamycin	_S. sp._	_(110)_

Table II. Examples of Other Non-_Streptomyces_ Microbial Compounds with Bioherbicidal Activity

Compound	Source	Reference
Acetylaranotin	_Aspergillus terreus_	_(122)_
Altersolanol A&B	_Alternaria porri_	_(123)_
Brefeldin	_A. carthami_	_(124)_
Citreoviridin	_Penicillium charlesii_	(125)
Colletotrichin	_Colletotrichins nicotianae_	_(126)_
Compactin	_Monascus sp_	_(127)_
Cyclopenin	_Penicillium cyclopium_	_(128)_
Dehydrocurvularin	_Alternaria macrospora_	_(129)_
Fusaric acid	_Fusarium oxysporum_	_(130)_
Mevinolin	_Aspergillus terreus_	_(131)_
Moniliformin	_Fusarium moniliforme_	_(132)_
Radicinin	_Alternaria helianthi_	_(133)_
Tentoxin	_A. alternata_	_(134)_
Viridiol	_Gliocladium virens_	(135)

XIII. Cornexistin   XIV. Cyperin

Other microbial bioherbicides.

studied for some time, nevertheless, most of these have not been (and will not be) directly developed as herbicides for a variety of reasons; e.g., non-target toxicity, difficulty and expense of synthesis, lack of high efficacy in field tests, etc. Furthermore, some compounds that will be found in the search for new phytotoxins from microbes and plants will be those that have been previously discovered. However, even when such phytotoxins fail to become viable commercial candidates, our knowledge of the relationship between molecular chemistry and phytotoxic activity is still enhanced. Some molecules that have been presented in this review have relatively simple chemical structures, while others are quite complex. These complex molecules are unlikely candidates for economic synthesis on a commercial scale, even if they possess desirable attributes. This could be a drawback, unless such compounds can be produced by large-scale fermentation. Another important aspect for such complex molecules is the discovery of their molecular modes of action. For naturally occurring compounds with unique target sites of action, quantitative structure:activity relationship (QSAR) protocols may allow the discovery of synthetic molecules with high phytotoxicity.

It is expected that herbicides, either synthetically derived or naturally occurring products, will remain a major method of weed control in order to produce food and fiber for the projected world population of 8 billion by the year 2025 (*141*). This need, and the success of obtaining phytotoxic compounds from microorganisms such as *Streptomyces,* suggest that the search for high potency phytotoxins from natural sources will continue. Less than 200,000 microbial species have been identified, but there may be up to 10-fold more yet unidentified, indicating an enormous reserve of potential sources for herbicidal products. Novel compounds with potential phytotoxicity from plants and microbes such as the *Streptomyces* and other species, will broaden our basic knowledge of chemistry as related to phytotoxic activity. Microbes that produce herbicidal compounds also possess the gene(s) that encode for the enzyme(s) that can metabolize or inactivate the phytotoxin. Such genes have potential use in the genetic transformation of crop plants for engineered resistance to the compound, which has been strikingly demonstrated in the case of phosphinothricin. Similar biotechnological approaches with other natural products that possess bioherbicidal activity will, no doubt, lead to the discovery of valuable agricultural products in the future.

References

1. Cutler, H.G. *Weed Technol.* **1988**, *2*, 525-532.
2. Duke, S.O.; Lydon, J. ACS Symp. Ser. **1993**, *524*, 110-124.
3. Duke, S.O.; Dayan, F.E.; Hernandez, A.; Duke, M.V.; Abbas, H.K. *Proc. Brighton Crop Protect. Conf.* **1997,** 579-586.
4. Fischer, H.-P.; Bellus, D. *Pestic. Sci.* **1983**, *14*, 334-346.
5. Hoagland, R.E. In *Microbial Products as Herbicides*; Hoagland, R.E., Ed.; ACS Symp. Ser. 439; Amer. Chem. Soc.: Washington, DC, 1990; pp 2-52.
6. *Microbial Products as Herbicides*; Hoagland, R.E., Ed.; ACS Symp. Ser. 439; Amer. Chem. Soc.: Washington, DC, 1990.
7. Hoagland, R.E. In *Advances in Microbial Biotechnology;* Tewari, J.,

Lakhanpal,T., Singh, J., Gupta, R., Camola, B., Eds.; APH Publishing Co. New Delhi,India, 1999; pp 213-255.

8. Hoagland, R.E.; Cutler, S.J. In *Allelopathy in Ecological Agriculture and Forestry*; Narwal, S.S., Hoagland, R.E., Dilday, R.H., Reigosa, M.J., Eds., Kluwer Academic Publishers, The Netherlands, 2000; pp 71-96.

9. Strobel, G.; Kenfield, D.; Bunkers, G.;Sugawara, F.; Clardy, J. *Experientia* **1991**,*47*, 819-826.

10. *Microbial Control of Weeds:* TeBeest, D. O.,Ed.; Chapman-Hall: New York, NY, 1991.

11. *Plant-Microbe Interactions and Biological Control*; Boland, G.J., Kuykendall, L.D., Eds.; Marcel Dekker, Inc.: New York,NY,1998, 442pp.

12. Klayman, D.L. *Science* **1985**, *228*, 1049-1055.

13. Bagchi, G.D.; Jain, D.C.; Kuman, S. *J. Medic. Arom. Plant Sci.* **1998**, *20*, 5-11.

14. Chen, P.K.; Leather, G.R. *J. Chem. Ecol.* **1990**,*16*, 1867-1876.

15. Duke, S.O.; Vaughan, K.C.; Croom, E.M., Jr.; Elsohly, H.N. *Weed Sci.* **1987**, *35*, 499-505.

16. Lori, H.S.; Leather, G.S.; Chen, P.K. *J. Chem. Ecol.* **1994**, *20*, 967-978.

17. Duke, M.V.; Paul, R.N.; Elsohly, H.N.; Sturtz, G.; Duke, S.O. *Internat. J. Plant Sci.* **1994**, *155*, 365-372.

18. DiTomaso, J.M.; Duke, S.O. *Pestic. Biochem. Physiol.* **1991**, *39*, 158-167.

19. Chen, P.K.; Leather, G.R. *J. Chem. Ecol.* **1990**, *16*, 1867-1876.

20. Muller, W.H.; Muller, C.H. *Bull. Torrey Bot. Club* **1964**, *91*, 327-330.

21. Grayson, B.T.; Williams, K.S.; Freehauf, P.A.; Pease, R.R.; Ziesel, W.T.; Sereno, R.L.; Reinsfelder, R.E. *Pestic. Sci.* **1987**, *21*, 143-153.

22. Lorber, P.; Muller, W.H. *Comp. Physiol. Ecol.* **1980**, *5*, 5-10.

23. El-Deek, M.H.; Hess, F.D. *Weed Sci.* **1986**, *34*, 684-688.

24. Romagni, J,G.; Duke, S.O.; Dayan, F.E. *Plant Physiol.* **2000**, (in press).

25. Carter, C.L.; McChesney, W.J. *Nature* **1949**, *166*, 575-576.

26. Hylin, J.W.; Matsumoto, H. *Arch. Biochem. Biophys.* **1960**, *93*, 542-545.

27. Kamikawa, T.; Higuchi, F.;Taniguchi, M.; Asaka, Y. *Agric. Chem. Biol.* **1980**, *44*, 691-692.

28. Hershenhorn, J.; Vurro, M.; Stierle, A.; Strobel, G. *Plant Sci.* **1993**, *94*, 227-234.

29. Hoagland, R.E. *Proc. South. Weed Sci. Soc.* **1995** , p 164.

30. Alston, T.A.; Mela, L. ; Bright, H.J. *Proc. Nat. Acad. Sci.,USA* **1977**, *74*, 37667-37671.

31. Coles, C.J..; Edmondson, D.E.; and Singer, T.P. *J. Biol. Chem.* **1979**, *54*, 5161-5167.

32. Netzley, D.H. ; Butler, L.G. *Crop Sci.* **1986**, *26*, 776-778.

33. Weston, L.A.; Nimbal, C.I. ; Czarnota, M.A. *Brighton Crop Protect.Conf.-Weeds.* **1997**, *6B-6*, 509-516.

34. Nimbal, C.I.; Yerkes, C.N.; Weston, L.A.; Weller, S.C. 1996, *Pestic. Biochem. Physiol.* **1997**, *54*, 73-83.

35. Gonzalez, V.M.; Kazimir, J.; Nimbal, C.; Weston, L.A.; Cheniae, G.M. *J. Agric. Food Chem.* **1997**, *45,* 1415-1421.

36. Nimbal, C.I.; Weston, L.A.; *Proc. 2nd Internatl. Weed Contr. Congr.* **1996,** *III*, 863-868.

37. Vaughn, K.C.; Vaughan, M.A. *Amer. Chem. Soc. Symp. Ser.* **1988**, *380*, 273-293.

38. Towers, G.H.N.; Arnason, J.T. *Weed Technol.* **1988**, *2*, 545-549.

39. Deshpande, B.S.; Ambedkar, S.S.; Shewale, J.G. *Enzyme Microb. Technol.* **1988**, *10*, 455-73.

40. Lavrik, P.B.; Isaac, B.G.; Ayer, S.W.; Stonard, R.J. *Dev. Ind.* **1991**, *32*, 79-91.

41. Stonard, R.J.; Miller-Wideman, M.A. In *Agrochemicals from Natural Products*; Godfrey, C.R.A., Ed.; Marcel Dekker Co.: New York, NY, 1995; pp. 285-310.

42. Tanaka, Y.; Omura, S. *Annu. Rev. Microbiol.* **1993**, *47*, 57-87.

43. Bayer, E.; Gugel, K.H.; Hägele, K.; Hagenmaier, H.; Jessipow, S.; König, W.A.; Zähner, H. *Helv. Chim. Acta* **1972**, *55*, 224-239.

44. Kondo, Y.; Shomura, T.; Ogawa, Y.; Tsuruoka, T.; Watanabe, H.; Totsukawa, K.; Suzuki, T.; Moriya, C.; Yoshida, J. *Sci. Rept. Meiji Seika Kaisha* **1973**, *13*, 34-44.

45. Rupp, W.; Finke, M.; Bieringer, H.; Langelueddeke, P. Germ. Offen. DE 2 717 440, **1977**, Hoechst AG.

46. Takematsu, T.; Konnai, M.; Tachibana, K.; Tsuruoka, T.; Inouye, S.; Watanabe, T. Germ. Offen. DE 2 848 224, **1979**, Meiji Seika Kaisha

47. Mase, S. *Jpn. Pestic. Inf.* **1984**, *45*, 27-30.

48. Deak, M.; Donn, G.; Feher, A.; Dudits, D. *Plant Cell Reports* **1988**, *7*, 158-161.

49. Wild, A.; Ziegler, C. *Z. Naturforsch.* **1989**, *44c*, 97-102.

50. Manderscheid, R.; Wild, A. *J. Plant Physiol.* **1986**, *123*, 135-142.

51. Vasil, I. K. 1996. Phosphinothricin-resistant crops. In *Herbicide- Resistant Crops,* Duke, S. O., Ed., pp. 85-91. Boca Raton, FL: CRC Press.

52. Dröge, W.; Broer, I.; Pühler, A. *Planta* **1992**, *187*, 142-151.

53. Murakami, T.; Anzai, H.; Imai, S.; Satoh, A.; Nagaska, K.; Thompson, C. *J. Mol. Gen. Genet.* **1986**, *205*, 42-50.

54. Kumada, Y.; Anzai, H.; Takano, E.; Murakami, T.; Hara, O.; Itoh, R.; Imai, S.; Satoh, A.; Nagaoka, K. *J. Antibiot.* **1988**, *41*, 1839-1845.

55. Thompson, C.J.; Movva, N.R.; Tigard, R.; Crameri, R.; Davies, J.E.; Lauwereys, S.M.; Botterman, J. *EMBO J.* **1987**, *6*, 2519-2523.

56. DeBlock, M.; Botterman, J.; Vanderwiele, M.; Dockz, J.; Thoen, C.; Gossele, V.; Movva, N.R.; Thompson, C.; van Montagu, J.; Leemans, *EMBO J.* **1987,** *6*, 2513-2518.

57. Broer, I.; Arnold, W.; Wohlleben, W.; Pühler, A. *Proc. Braunschweig. Symp. Applied Plant Molec. Biol.* **1989**, 240-250.

58. Wohlleben, W.; Arnold, W.; Broer, I.; Hillemann ,D.; Strauch, E.; Pühler, A. *Gene* **1988**, *70*, 25-37.

59. Liu, C. A.; Zhong, H.; Vargas, J.; Penner, D.; Sticklen, M. *Weed Sci.* **1998**, *46*, 139-146.

60. De Datta, S.K. 1981. *Principles and Practices of Rice Production.* New York: Wiley International.

61. Kondo, Y.; Shomura, T.; Ogawa, Y.; Tsuruoka, T.; Watanabe, H.; Totsukawa, K.; Suzuki, T.; Moriya, C.; Yoshida, J. *Sci. Rept. Meiji Seika Kaisha* **1973**, *13*, 34-44.

62. Uchimiya, H.; Iwata, M.; Nojiri, C.; Samarajeewa, P. K.; Takamatsu, S.; Ooba, S.; Anzai, H.; Christensen, A.H.; Quail, P.H.; Toki, S. *Bio/Technology* **1993**, *11*, 35-836.

63. Tada, T.; Kanzaki, H.; Norita, E.; Uchimiya, H.; Nakamura, I. *Molec. Plant-Microbe Interact.* **1996**, *9*, 762-764.

64. Kumada, Y.; Imai, S.; Nagoaka, K. *J. Antibiot.* **1991**, *44*, 1006-1012.

65. Omura, S.; Hinotozawa, T.; Imamura, N.; Marata, M. *J. Antibiot.* **1984**, *37*, 939-940.

66. Omura, S.; Murata, M.; Hanaki, H.; Hinotozawa, T.; Oiwa, R.; Tanaka, H. *J. Antibiot.* **1984**, *37*, 829-835.

67. Kato, H.; Nagayana, K.; Abe, H.; Kobayashi, R.; Ishihara, E. *Agric. Biol. Chem.* **1991**, *55*, 1133-1134.

68. Hoagland, R.E. In *Biologically Active Natural Products in Agriculture and Pharmaceuticals*; Cutler, H.G. and Cutler, S. J., Eds.; CRC Press; Boca Raton, FL, 1999; pp 107-125.

69. Tate, S.S.; Meister, A. *The Enzymes of Glutamine Metabolism*; Academic Press, New York, NY, 1973; pp 77-86.

70. Jeannoda,V.L.; Valisolaloa, J.; Creppy, E.E.; Dirheimer, G. *Phytochemistry* **1985**, *24*, 854-855.

71. Gass, J.D.; Meister, A. *Biochemistry* **1970**, *9*, 1380-1390.

72. Manning, J.M.; Moore, S.; Rowe, W. B.; Meister, A. *Biochemistry* **1969**, *8*, 2681- 2685.

73. Ronzio, R.A.; Rowe, W.B.; Meister, A. *Biochemistry* **1969**, *8*, 1066-1075.

74. Sekizawa, Y.; Takematsu, T. In *Pesticide Chemistry, Human Welfare and the Environment, Vol. 2, Natural Products*; Pergamon Press: Oxford, England, 1983; pp 261-68.

75. Pruess, D.L.; Scannell, J.P.; Ax, H.A.; Kellett, M.; Weiss, F.; Demny, T.C.; Stempel, A. *J. Antibiot.* **1973**, *26*, 261-266.

76. Scannell, J.P.; Pruess, D.L.; Demney, T.C.; Ax, H. A.; Weiss, F.; Williams, T.; Stempel, A. In *Chemistry and Biology of Peptides;* Watler, R. and Meinhafer, J. Eds.; Ann Arbor Science Publishers : Ann Arbor, MI, 1972; pp 415-421.

77. Walworth, B. L. U.S. Patents 3 295 949 and 3 323 895, 1967.

78. Wild, A.; Mandersheid, R. *Z. Naturforsch.* **1994**, *39c*, 500-504.

79. Lea, P.J.; Ridley, S.M. In *Herbicides and Plant Metabolism*; Dodge, A.D Ed.; Soc. Exp. Biol. Seminar Ser. No. 38; Cambridge Univ. Press: Cambridge, England, 1989; pp. 137-170.

80. Logusch, E.W.; Walker, D.M.; McDonald, J.F.; Franz, J.E. *Plant Physiol.* **1991**, *95*, 1057-1062.

81. Logusch, E.W.; Walker, D.M.; McDonald, J.F.; Franz, J.E.; Willafranca, J.J.; DiIanni, C.L.; Colanduoni, J.A.; Li, B.; Schineller, J.B. *Biochemistry* **1990**, *29*, 366-372.

88

82. Mastalerz, P. *Rocz. Chem.* **1959**, *33*, 985-991.
83. Mastalerz, P. *Arch. Immunol. I. Terapii Doświeadczalnej* **1959**, 7, 201-210.
84. Yamada, O.; Kaise, Y.; Futatsuya, F.; Ishida, S.; Ito, K.; Yamamoto, H.; Munakata, K. *Agric. Biol. Chem.* **1972**, *36*, 2013-2015.
85. Felner, I.; Schenker, K. *Helv. Chim. Acta* **1970**, *53*, 754-763.
86. The Pesticide Manual, 6th Edition. 1979. British Crop Protection Council, Croydon; p. 349.
87. Rupp, W.; Finke, M.; Bieringer, H.; Langelueddeke, P. Germ. Offen. DE 2 717 440, **1977**, Hoechst AG.
88. Arai, M.; Haneishi, T.; Kitahara, N.; Enokita, R.; Kawakubo, K.; Kondo, Y. *J. Antibiot.* **1976**, *29*, 863-869.
89. Isaac, B.G.; Ayer, S.W.; Elliott, R.C.; Stonard, R.J. *J. Org. Chem.* **1992**, *57*, 7220-7226.
90. Miller-Wideman, M.; Makkar, N.; Tran, M.; Isaac, B.G.; Biest, N. *J. Antibiot.* **1992**, *45*, 914- 921.
91. Edmunds, A.J.; Trueb,W.; Oppolzer, W.; Cowley, P. *Tetrahedron* **1997**, *53*, 2785-2802.
92. Takahashi, S. *Amer. Chem. Soc. Symp. Ser.* **1994**, *551*, 74-84.
93. Mio, S.; Ichinose, R.; Goto, K.; Sugai, S. *Tetrahedron* **1991**, *47*, 2111-2120.
94. Mio, S.; Shiruishi, M.; Sugai, S.; Haruyama, H.; Sato, S. *Tetrahedron* **1991**, *47*, 2121-2132.
95. Mio, S.; Ueda, M.; Hamura, M.; Kitagawa, J.; Sugai, S. *Tetrahedron* **1991**, *47*, 2145-2154.
96. Nakajima, M.; Itoi, K.; Takamatsu, Y.; Kinoshita, T.; Okazaki, T. *J. Antibiot.* **1991**, *44*, 293-300.
97. Fonne-Pfister, R.; Chemla, P.; Ward, E.; Girardet, M.; Kreuz, K.E.; Honzatko, R.B.; Fromm, H.J.; Schar, H.P.; Grutter, M.G.; Cowan-Jacob, S.W. *Proc. Natl. Acad. Sci., USA* **1996**, *93*, 9431-9436.
98. Heim, D.R.; Cseke, C.; Gerwick, B.C.; Murdoch, M.G.; Green, S.B. *Pestic. Biochem. Physiol.* **1995**, *53*, 138-145.
99. Siehl, D.L.; Subramanian, M.V.; Walters, E. W.; Lee, S.-F.; Anderson, R.J.; Toshi, A.G. *Plant Physiol.* **1996**, *110*, 753-758.
100. Poland, B.W.; Lee, S.-F.; Subramanian, M.V.; Siehl, D.L.; Anderson, R.J.; Fromm, H.J.; Honzatko, R.B. *Biochemistry* **1996**, *35*, 5753-5759.
101. Gerwick, B.C.; Fields, S.S.; Graupner, P.R.; Gray, J.A. *Weed Sci.* **1997**, *45*, 654-657.
102. Lavrik, P.; Isaac, B.; Ayer, S.; Stonard, R.; *Dev. Ind.. Microbiol.* **1991**, *32*, 79-91.
103. Norman, A.; *Science* **1955**, *121*, 213-214.
104. Issac, B .; Ayer, S.; Letendre, L.; Stonard, R. J.; *J. Antibiot.* **1991,** *44*, 729-732.
105. Kida, T.; Shibai, H.; *Agric. Biol. Chem.* **1986**, *50*, 483-484.
106. Waller, V.; Bell, W.; *Plant Dis. Rept.* **1956**, *40*, 129-132.
107. Murao, S.; Hayashi, H. *Agric. Biol. Chem.* **1983**, *47*, 1135-1136.

108. Kobinata, K.; Sekido, S.; Uramoto, M.; Ubukata, M.; Osada, H.; *Agric. Biol. Chem.* **1991**, *55*, 1415-1416.
109. Berg, D.; Schedel, M.; Schmidt, R.; Ditgens, K.; Weyland, H.; *Z. Naturforsch.* **1982**, *37c*, 1100- 1106.
110. Yamada, O.; Kaise, Y.; Futatsuya, F.; Ishida, S.; Ito, K.; Yamamoto, H.; Munakata, K. ; *Agric. Biol. Chem.* **1972**, *36*, 2013-2015.
111. Fushimi, S.; Nishikawa, S.; Mito, N.; Ikemoto, M.; Sasaki, M.; Seto, H. *J. Antibiot.* **1989**, *42*, 1370-1378.
112. Kristinsson, H.; Nebel, K.; O'Sullivan, A.C.; Pachlatko, J.P.; Yamaguchi ,Y. *Amer. Chem. Soc. Symp. Ser.* **1995**, *584*, 206-219.
113. Babczinski, P.; Dorgerloh, M.; Löbberding, A.; Santel, H.-J.; Schmidt, R.R.; Schmitt, P.; Wünsche, C. *Pestic. Sci.* **1991**, *33*, 439-446.
114. Tanaka, Y.; Sugoh, M.; Yoshida, H.; Arai, N.; Shiomi, K.; Matsumoto, A.; Takahaski, Y.; Omura, S. *J. Antibiot.* **1995**, *48*, 1525-1526.
115. Nakajima, M.; Itoi, K,; Takamatsu, Y.; Sato, S.; Furukawa, Y.; Furuya, K.; Honma,T.; Kotani, Y.; Kozasa, M.; Haneishi, T. *J Antiobiot.* **1989**, *44*, 1065-1072.
116. Amagasa, T.; Paul, R.N.; Heitholt, J.J.; Duke, S.O. *Pestic. Biochem. Physiol.* **1994**, *49,* 37-52.
117. Nishino, T,; Murao, S. *Agric. Biol. Chem.* **1983,** *47*, 1961-1966.
118. Weber, H.A.; Gloer, J.B. *J. Nat. Prod.* **1988**, *51*, 897-883.
119. Stierle, A.; Upadahyay, R.; Strobel, G. *Phytochemistry* **1991**, *30*, 2191-2192.
120. Venkatasubbaiah, P.; Van Dyke, C.G.; Chilton, W.S. *Mycologia* **1992**, *84*, 715-723.
121. Harrington, P.M.; Singh, B.K.; Szamosi, I.T.; Birk, J.H. *J. Agric. Food Chem.* **1995**, *43*, 804-808.
122. Kamata, S.; Sakai, H.; Hirota, A.; *Agric. Biol. Chem.* **1983**, *47*, 2637-2638.
123. Suemitsu, R.; Yamada, Y.; Sano, T.; Yamashita, K.; *Agric. Biol. Chem.* **1984**, *48*, 2383-2384.
124. Tietjen, K.; Schaller, E.; Matern, U.; *Physiol. Plant Pathol.* **1983**, *23*, 387-400.
125. Cole, R.; Dorner, J.; Cox, R.; Hill, R.; Cutler, H.; Wells, J.; *Appl. Environ. Microbiol.* **1981**, *42*, 677-681.
126. Gohbara, M.; Kosuge, Y.; Yamasaki, S.; Kimura, Y.; Tamura, S.; *Agric. Biol. Chem.* **1978**, *42*, 1037-1043.
127. Hashizume, T.; *Agric. Biol. Chem.* **1983***, 47*, 1401-1403.
128. Cutler, H.; Crumley, F.; Cox, R.; Wells, J.; Cole, R.; *Plant Cell Physiol.* **1984**, *25*, 257-263.
129. Robeson, D. ; Strobel, G.; *J. Nat. Prod.* **1985**, *48*, 139-141.
130. Chakrabarti, D.; Chaudhury, K.; Ghosal, S; *Experientia* **1976**, *32*, 608-609.
131. Bach, T.J.; Lichtenthaler, H.K. *Physiol. Plant.* **1983**, *59*, 50-60.
132. Cole, R.; Kirksey, J.; Cutler, H.; Doupnik, B.; Peckham, K.; *Science* **1973**, *179*, 1324-1326.
133. Tal, B.; Robeson, J.; Burke, B.; Aasen, A.; *Phytochemistry* **1985**, *24*, 729-731.
134. Lax, A.; Shepard, H.; Edwards, J.; *Weed Technol.* **1988**, *2*, 540-544.
135. Howell, C.; Stipanovic, R.; *Phytopathology* **1984***, 74*, 1346-1349.

136. Bach, T.J.; Lichtenthaler, H.K. In *Ecology and Metabolism of Plant Lipids*; Fuer, G. and Nes, W. D., Eds.; *Amer. Chem. Soc. Symp. Ser.* **1987**, *235,* 109-139.

137. Josekutty, P.C. *S. Afr. J. Bot.* **1998**, *64*, 18-24.

138. Yamazaki, S.; Okuba, A.; Akiyama, Y.; Fuwa, K. *Agric. Biol. Chem.* **1975**, *39*, 287-289.

139. Youngman, R.J.; Elstner, E.F. In *Oxygen Radicals in Chemistry and Biology*; Bors, W., Saran, M.; Tait, D.; Eds. Walter deGruyter, New York, NY.; 1984, p 501-507.

140. Hoagland, R.E. *Weed Technol.* **1996**, *10*, 651-674.

141. United States Bureau of the Census. 1998. Total midyear population for the world:1950-2050. http://www.census.gov/ipc/www/worldpop.html

Chapter 8

Conjugated and Unconjugated Brassinosteroids

Hiroshi Abe, Kazuo Soeno, Naoko-N Koseki, and Masahiro Natsume

**Department of Applied Biological Science, Tokyo University
of Agriculture and Technology, Fuchu, Tokyo, 183–8509, Japan**

Research on brassinosteroids(BRs) started with the discovery of
Distylium factors. Brassin technology is growing rapidly and
advancing in the knowledge of biosynthesis and metabolism,
and in molecular genetic analysis using BR-deficient mutants.
BR is an essential hormone for regulation of plant growth and
development. BR can be regarded as the most important
discovery in the field of plant growth regulation. BR will be
made avilable for increasing crop production and crop protection
as a new type of plant growth regulator in the near future.

Among new class of plant hormones, brassinosteroid(BR) is a
steroidal compound with a unique chemical structure and specific biological
activity. More than 40 kinds of BR have been identified in plants, ferns and
algae but it has not yet been found as a microbial metabolite. Natural BR
occurs in either an uncojugated or conjugated form, and the former is both
in abundance and ubiquitous in plants. Two biosynthetic routes for
brassinolide(BL), a typical and the most active BR in nature, have been
discovered. A variety of modified BRs have been also detected in metabolic
studies using plant seedlings or plant cell cultures, including epimerization,
hydroxylation, fatty acid esterification, and glycosylation. BR elicits

growth promotion, stress tolerance enhancement and crop yield increase, indicating that it will be available as a new type of plant growth regulator for increasing crop production and crop protection. In this paper we describe the occurrence of conjugated and unconjugated BRs in nature, the biosynthetic and metabolic pathways of BRs, and the prospect and problems for agricultural application.

Unconjugated Brassinosteroids

Natural BRs were purified from plant tissue through monitoring elongation or splitting activity in the bean first or second internode test and promotion activity in the rice-lamina inclination test. General purification procedure in our laboratory consisted of 5 steps; solvent extraction, acetonitrile-*n*-hexane partition, SiO_2 or Al_2O_3 adsorption chromatography, reversed-phase Sep-Pak cartridge column chromatography, and finally C_{18} reversed-phase HPLC. The purified active fraction was subjected to GC/MS analysis after conversion into methaneboronate or methaneboronate-TMS, or subjected to LC/FABMS analysis without derivatization. Accordingly, more than forty compounds have been isolated and identified. However, many compounds remain unidentified, so that the number of identified BRs is likely to increase.

The structure-activity relationship of naturally occurring BRs and related synthetic compounds were studied by means of the lamina inclination test. The results indicated that BR activity required two pairs of vicinal diols at ($2\alpha, 3\alpha$) and (22R,23R), B-ring of 7-oxalactone, one alkyl substituent at the C24 position, and the A/B-trans ring junction.

The results of structural variation and their biological activity relationship proposed two hypothetical biosynthetic pathways for BL from campesterol. Metabolic studies in culture cells of *Catharanthus roseus* and its transformed cells, and biosynthetic mutants of *Arabidopsis* and *Pisum sativum* revealed that BL is synthesized from campesterol by either early C6-oxidation pathway or late C6-oxidation pathway as shown in Figure 1(*1*). However, 2-epiCastasterone(2-epiCS)(**1**), 3-epiCS(**2**), 2,3-diepiCS(**3**) and $2\beta, 3\beta$-epoxyCS (**4**, secasterone) were not involved in the two pathways, suggesting that another biosynthesis pathway for these compounds may be involved in plants.

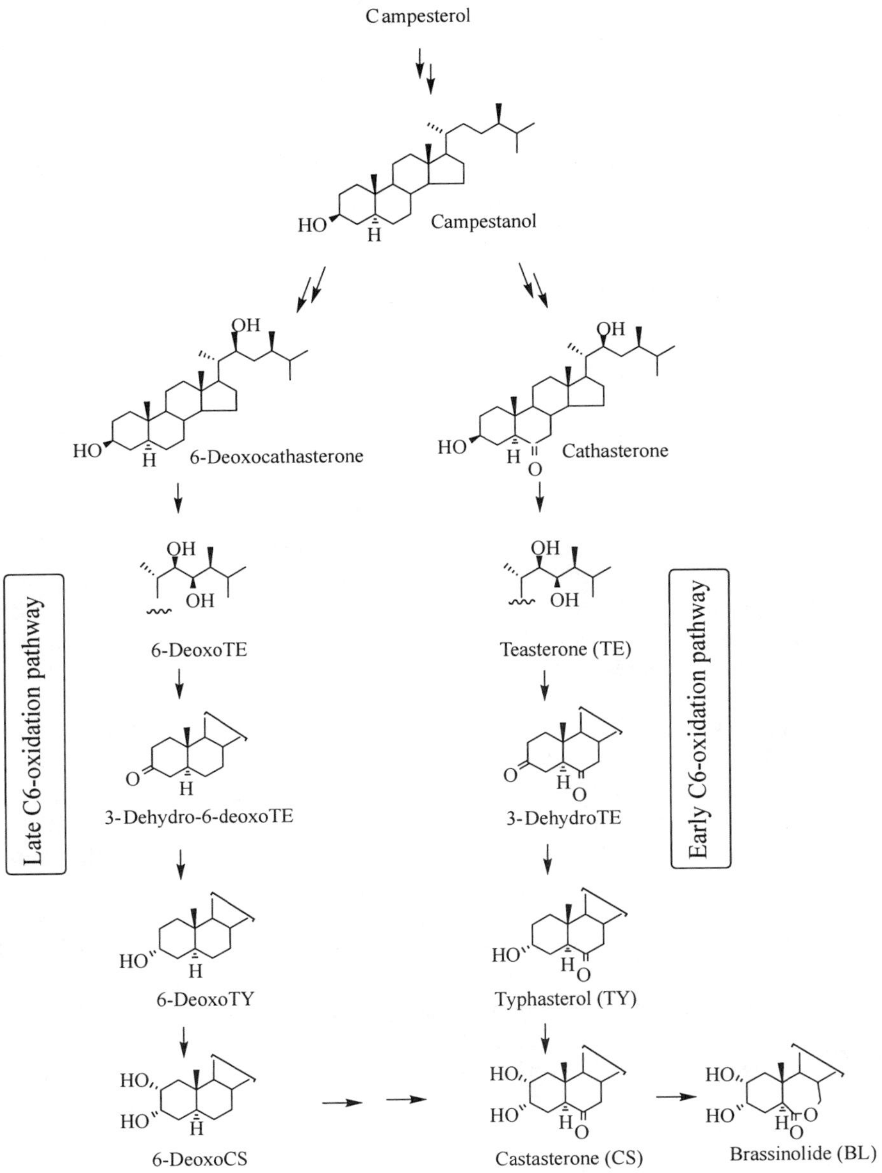

Figure 1 Biosynthetic pathways of brassinolide from campesterol

(1) (2) (3) (4)

Conjugated Brassinosterids

Conjugation of BRs has been investigated intensively during the past 5 years and good progress was made concerning their chemistry and physiological significance. Conjugated BRs have been found as endogenous BR and metabolites converted from exogenously applied BRs. 25-Methyl dolichosterone-23 β-D-glucoside(**5**) and its 2 β isomer(**6**) from *Phaseolus vulgaris* seeds(*2*), and teasterone-3 β-D-glucoside(TE-3 β-D-glucoside)(**7**) (*3*), TE-3-laurate (**18**) and TE-3-myristate (**19**)(*4*) from *Lilium longiflorum* pollen have been identified as endogenous BRs (Figures 2, 3).

In addition to the endogenous conjugates, a variety of conjugated BRs have been found in metabolic studies including glycosylation, esterification and acylglycosylation. Glucosyl conjugation has been observed in metabolic studies (Figure 2). 23 β-D-glucoside (**8**)(*5*) was found in BL metabolism using mung bean explant. 25 β-D-Glucoside (**9**) and 26 β-D-glucoside (**10**)(*6*) were major metabolites of 24-epiBL in tomato culture cells. This metabolism indicated that hydroxylation at C25 and C26 was performed prior to glucosylation. The same conjugation pathway has been observed in the metabolism of 24-epiCS, using tomato culture cells, which

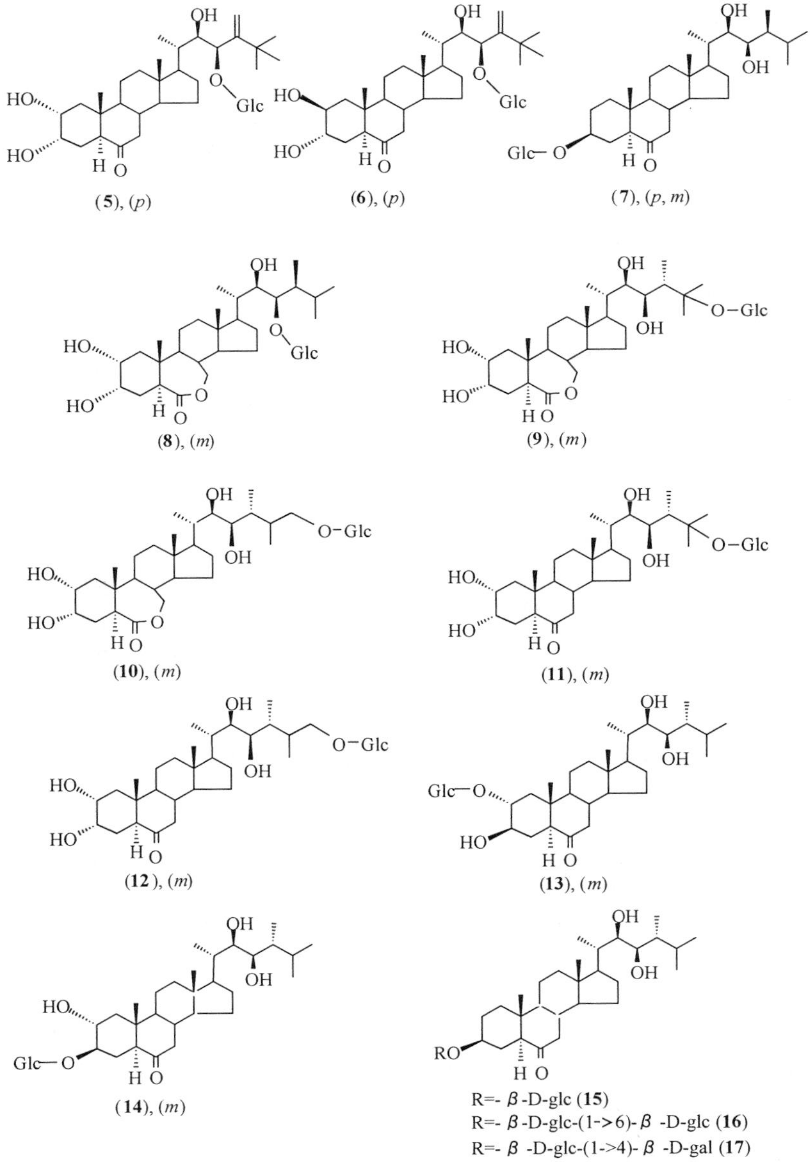

Figure 2 Structures of glycosylated brassinosteroid conjugates detected endogenously(p) and in metabolism(m).

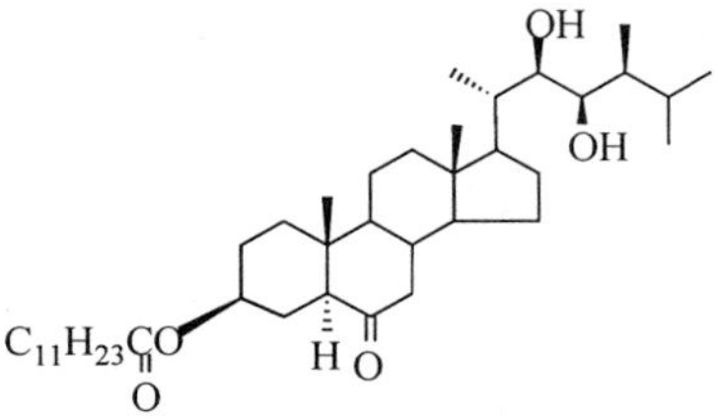

(**18**) (*p*, *m*)

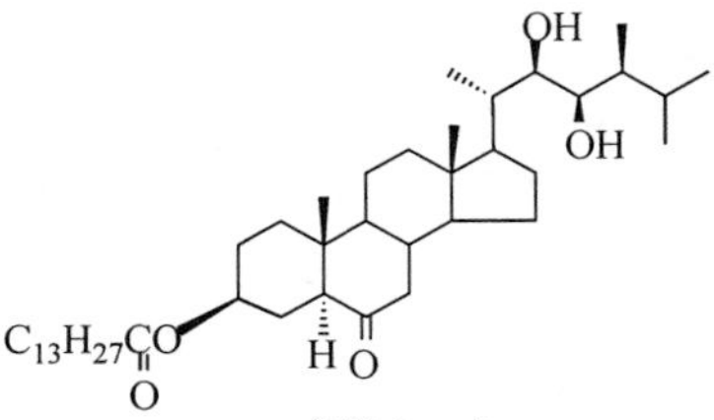

(**19**) (*p*, *m*)

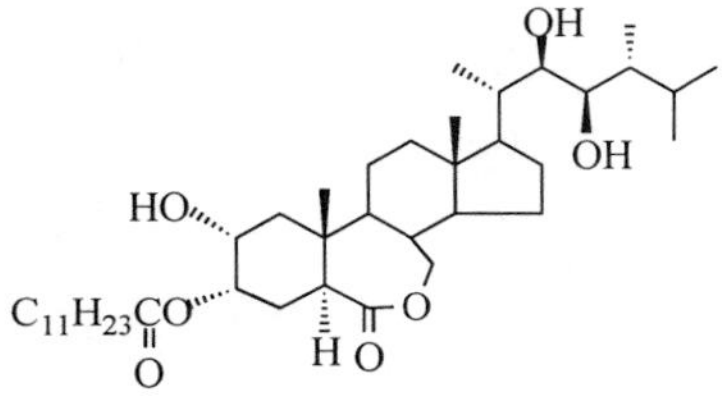

(**20**) (*m*)

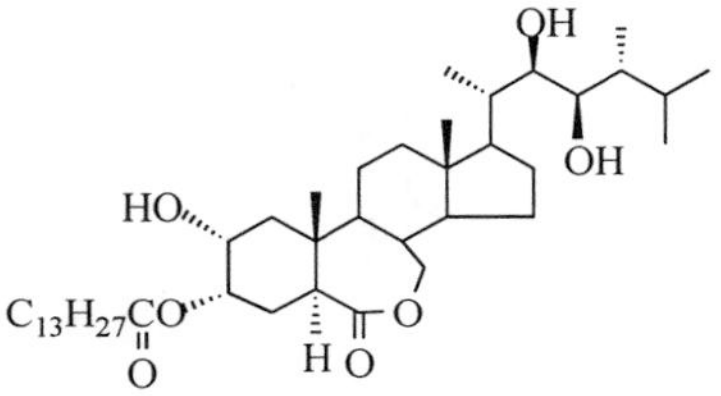

(**21**) (*m*)

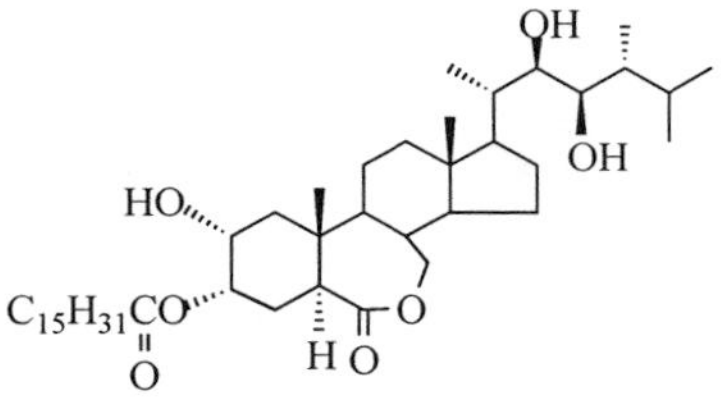

(**22**) (*m*)

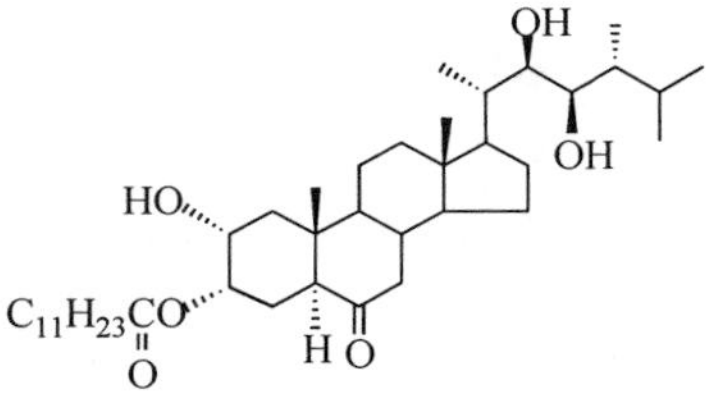

(**23**) (*m*)

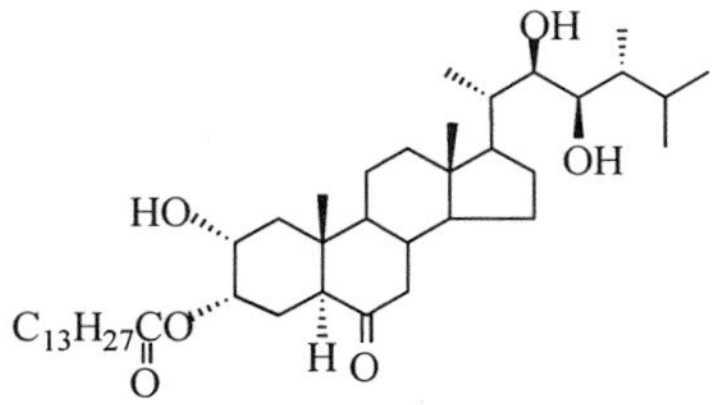

(**24**) (*m*)

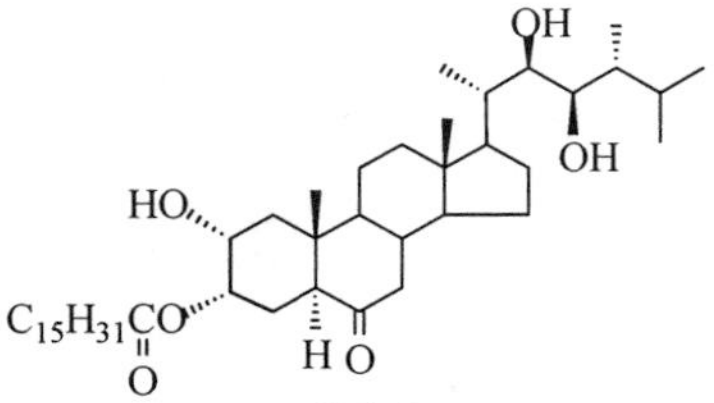

(**25**) (*m*)

Figure 3 Structures of esterified brassinosteroid conjugates detected endogenously(p) and in metabolism(m).

was converted into 25 β-D-glucoside(11) and 26 β-D-glucoside(12)(6). However, the metabolism of 24-epiCS by tomato culture cells also produced 2 β-(13) and 3 β-glucosides(14)(6) in addition to the side chain glucosides. The results indicated that α to β epimerization of the corresponding 3-hydroxyl was important prior to the formation of the glucosyl conjugation. Requirement of epimerization prior to the conjugation has been observed at the C2- or C3-hydroxyl in the metabolism of BL and its 24-epimer by cucumber seedlings or in culture cells(7), although the structure elucidation remains to be determined. Glucosylation was observed in the metabolism of TE. TE-glucoside(7) was found in metabolites of TE afforded by meristem cell cultures of *Lilium longiflorum*(3). 24-epiTE-3 β-D-glucoside(15), its (1-6)- and (1-4)-3 β-D-glucosides(16,17) have been obtained from exogenously applied 24-epiTE in cell cultures of *Lycopersicon esculentum* L.(8).

On the other hand, esterification has been observed in the metabolism of BR(Figure 3). 3 β-Lauryl(20), 3 β-myristyl(21) and 3 β-palmityl(22) derivatives were obtained as metabolites of 24-epiBL by *Ornithopus sativus* culture cells(6). The formation of the ester conjugation (23,24,25)(6) required α to β epimerization of the 3-hydroxyl prior to the esterification in the metabolism of 24-epiCS by tomato culture cells as shown in the case of glucosylation. Exogenously applied TE was converted to TE esters (18) and (19) by *Lilium longiflorum* cell cultures which were completely compatible with endogenous esters in the pollen (9). Metabolic studies pointed out that many kinds of metabolic pathways were found in plants and the pathway depended on plant species, plant organs and substrate.

The formation of conjugation and its physiological significance were investigated through metabolism of TE in the cell cultures and pollen growth of *Lilium longiflorum*. The pollen contained campesterol, TE, 3-dehydroTE, typhasterol(TY), CS and BL, postulating that BL was biosynthesized *via* the early C6-oxidation pathway from campesterol (Figure 1). Only TE was detected in the esterified (18,19) and glucosyl (7) conjugated forms among the endogenous BRs. In the time course fluctuation of TE and TE-ester during pollen growth, a maximum concentration of the endogenous TE-ester appeared in the immature pollen stage but that of TE only in the mature pollen stage. During the time lag between the immature stage and the mature stage, the accumulated TE-ester must be hydrolyzed to release free TE and then converted into an active brassinosteroid, either CS or BL. It strongly suggests that the conjugation acts as a reversible deactivation storage form for the biosynthetic pathway of BL.

In order to examine the reversible reaction between TE and TE-ester,

98

[28-^{14}C]-labeled TE and its myristate were prepared and metabolites, after feeding to the culture cells, were identified. TE was converted to 3 β-laurate(**18**) and 3 β-myrstate(**19**), respectively. TE-myristate was easily converted to free TE after 24h incubation. These results indicated that the ester conjugation was a reversible reaction. The same reversible reaction was observed between TE and its glucoside (**7**).

These results concluded that (1) Conjugation of TE is a reversible deactivation reaction, (2) Ester or glycosyl conjugation may act as a storage or transport component. The conjugates are hydrolyzed for BL biosynthesis (Figure 4).

Prospect for Agricultural Application

There is no doubt that BRs play a regulatory role in plant growth and development. Plant growth acceleration, crop yield increase, and increase of stress tolerance to cold temperature, salt, disease and herbicidal injury are noticeable effects for agricultural application. However, there are some problems with the agricultural application, that is, high cost, short duration, and restricted effect.

We examined the metabolism of [28-^{3}H]-BL in cucumber seedlings. The roots were pulse labeled for 6h, washed with distilled water, and then chased for 18h. The 1st leaves, epicotyls, cotyledons and hypocotyls were harvested. Solvent extracts were separated into the *n*-hexane-soluble conjugate fraction (less polar metabolites), the water-soluble conjugate fraction (polar metabolites) and the ethyl acetate-soluble fraction (unconjugated metabolites). Radioactivity was detected in all fractions. Hydrolysate of the less polar and the polar metabolite fractions gave a BL epimer, which differed from BL in retention time on LC-frit-FABMS, but its mass spectrum was quite similar to that of BL. GC/MS spectrum of the epimer indicated that it should be a streoisomer of the hydroxyl on the A-ring although the structural elucidation remains to be clarified. Exogenously treated BL was quickly metabolized to the stereoisomeric epimer and then the epimer was immediately transformed into its conjugated form binding with glucose and/or fatty acid. The conjugation reaction was estimated to be an irreversible deactivation process. As mentioned above, epimerization of the 3 α-hydroxyl prior to conjugation was observed in the metabolic study. It is important to develop a compound so that such epimerization or conjugation reaction dose not occurr in treated plant tissue. Several synthetic compounds (**26,27,28**) have been made available for testing *in vitro* and *in vivo*.

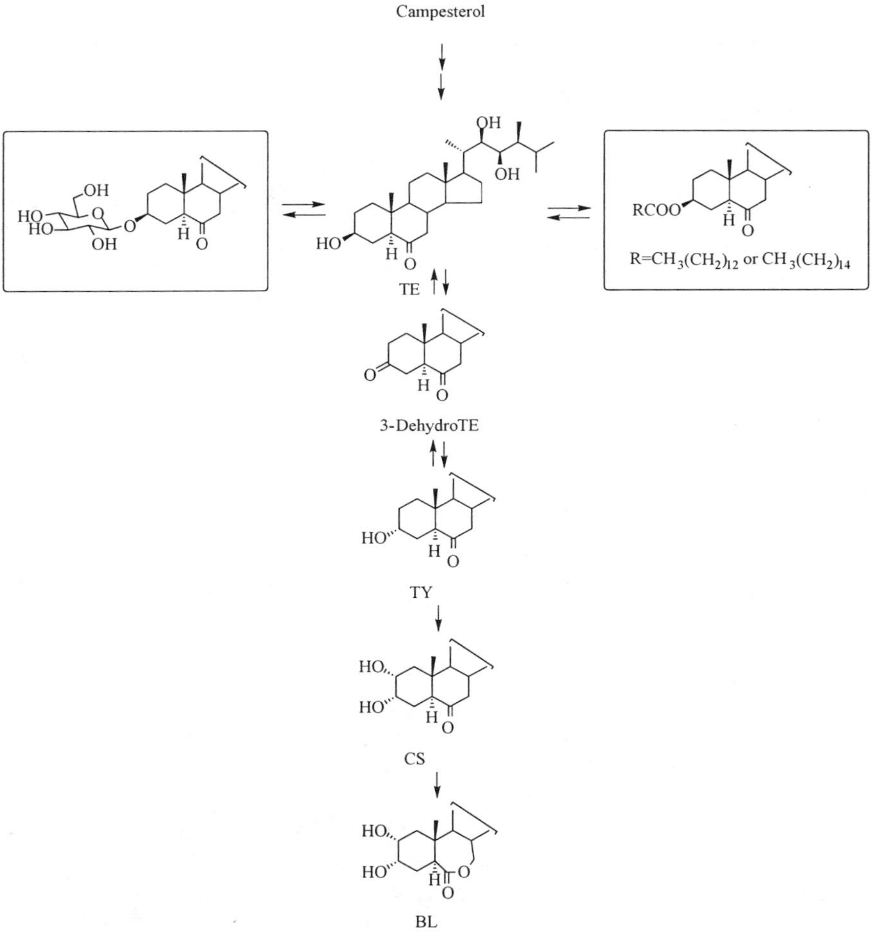

Figure 4   Biosynthesis and metabolism of unconjugated and conjugated brassinosteroids in both the pollen and the culture cells of Lilium longiflorum.

Regarding the agricultural application of BRs, there are a few interesting studies on synergistic or antagonistic compounds. Synergistic effects of 3 β -glucosides of phytosterol, stigmasterol, β -sitosterol and campesterol were tested in rice-lamina inclination assays and mung bean for internode elongation activities. Synergistic effect of BL was observed only with stigmasteryl-3-*O*-β-D-glucoside(**29**) but not with β-sitosterylglucoside and campesteryl glucoside(*10*). More recently, a biosynthetic inhibitor, named brassinazole(**30**), which blocks P450 enzyme has been found by Min *et al.*(*11*) at RIKEN in Japan (see the details in Chapter reported by Yoshida).

(26)   (27)

(28)   (29)   (30)

Discussion

Two independent studies, the Brassin project by the USDA group in the USA and *Distylium* factors by the Nagoya University group in Japan initiated BR research. Brassin was identified to be BL and its discovery led to the successful isolation and identification of all other homologues in the plant kingdom using the rice-lamina inclination test. In addition to the chemical and physiological studies in BR research, recent remarkable advances in BR research are revealed in the molecular identification and characterization of BR-deficient and BR-insensitive mutants; *Arabidopsis*, *Pisum sativum* and tomato, and the corresponding genes. BR was confirmed to be an essential hormone that plays a regulatory role in plant growth and development. We anticipate that the next development will be revelation of the physiological function and mode of action of endogenous BR in plant growth and development.

BR may be regarded as the most important discovery in the field of plant growth regulation. Nobody doubts that BR will be made available for crop production and crop protection as a new type of plant growth regulator in the near future.

References

1. Yokota, T. In *Biochemistry and Molecular Biology of Plant Hormones*; Hookaas, P. J. J., Hall, M. A., Libbenga, K. R.,Eds.; Elsevier Science B.V., Amsterdam, The Netherland, 1999, pp.277-293.
2. Yokota, T.; Kim, S. K.; Kosaka, Y.; Ogino, Y.; Takahashi, N. In Conjugated Plant Hormnones: Structure, Metabolism and Function.; Schreiber, K.; Schuter, H.R.; Sembdner, G., Eds.; VEB Deutscher Verlag der Wissenschaften, Berlin, 1987, pp.288-296.
3. Soeno, K.; Kyokawa, Y.; Natsume, M.; Abe, H. Biosci. Biochem. Biotech. 2000, 65, in press.
4. Asakawa, S.; Abe, H.; Nishikawa, N.; Natsume, M.; Koshioka, M. Biosci. Biochem. Biotech. 1996, 60, 1416-1420.
5. Suzuki, H.; Kim, S.K.; Takahashi, N.; Yokota, T Phytochemistry 1993, 33, 1361-1367.
6. Kolbe, A.; Schneider, B.; Porzel, A.; Schmidt, J.; Adam, G. Phytochemistry 1995, 38, 633-636.
7. Nishikawa, N.; Abe, H.; Natsume, M,; Shida, A.; Toyama, S. J.Plant Physiol. 1995, 147, 294-300.
8. Kolbe, A.; Porzel, A.; Schneider, B.; Adam, G. Phytochemistry 1997, 46, 1019-1022.
9. Soeno, K.; Asakawa, S.; Natsume, M,; Abe, H. J. Pest. Sci. 2000, 25, in press.
10. Kyokawa, Y.; Abe, H.; Natsume, M.; Koshioka, M. J. Pest. Sci. 1996, 21, 209-211.
11. Min, Y. K.; Asami, T.; Fujioka, S.; Yamaguchi, I.; Murofushi, N.; Yoshida, S. Bioorg. Med. Chem. Lett. 1999, 9, 425-430.

Chapter 9

Bioactive Substances from Medicinal Plants

A. Douglas Kinghorn[1], Leng Chee Chang[1], and Baoliang Cui[1, 2]

[1]Program for Collaborative Research in the Pharmaceutical Sciences
and Department of Medicinal Chemistry and Pharmacognosy,
College of Pharmacy, University of Illinois at Chicago, Chicago, IL 60612
[2]Current address: Phytera, Inc., Worcester, MA 01605

Higher plants produce a plethora of bioactive secondary
metabolites, with these compounds thought to serve primarily
as defensive substances against animal and plant predators. In
this chapter, several known and novel compounds are
described from five medicinal plants, with particular emphasis
made on their potential use as cancer chemotherapeutic agents
or cancer chemopreventives. Reference is made to the
previously documented pesticidal activities of these species
and their constituents.

Introduction

At first glance, the inclusion of a chapter on bioactive principles from
medicinal plants in a volume on the chemistry of natural product pesticides
would seem be somewhat superfluous. However, when the compound classes
included in a previous symposium volume on biologically active natural
products of potential use in agriculture (*1*) are compared with those in a similar
volume on human medicinal agents from plants (*2*), considerable overlap is

apparent. For example, the quinoline alkaloid antitumor agent camptothecin has been shown to be a seed germination inhibitor for various crop plants (*2,3*), while the promising sesquiterpene antimalarial artemisinin has proven to be a potent phytotoxin (*2,4*). Along similar lines, glucosinolates produced by various *Brassica* species are of interest not only because of their demonstrated pesticidal effects on insects, and fungi, and as allelopaths (*5*), but also because of their potential cancer chemopreventive effects, as exhibited by the induction of phase II enzymes relevant to the metabolism and detoxification of carcinogens and since they inhibit the formation of tumors in certain mammalian *in vivo* test systems (*6*).

Plant natural chemicals having potential use as either pharmaceuticals, food additives, pesticides, allelochemicals, or plant-growth regulators tend to be high-value secondary metabolites (*7*). While there is some debate as to the exact physiological role of secondary metabolites in the plant, it has been hypothesized that secondary metabolites have evolved to enhance the survival of the producing organism by having the ability to act at specific receptors in competing organisms (*8*). Thus, such secondary metabolites, which are compounds less than 3,000 Daltons in molecular weight, are regarded as having an ecological role, with no apparent role in the primary metabolism of the plant (*7,8*). The common properties of plant secondary metabolites and their biological cross-reactivity in research programs oriented either to pharmaceutical drug discovery or to pesticide discovery has been noted by others (*9*).

In this chapter, examples of novel and/or biologically active compounds are presented from two ongoing phytochemical research projects, directed towards the discovery of antitumor agents (*10,11*) and cancer chemopreventives (*12,13*), respectively. Plants in these two projects are obtained especially from tropical rainforest regions of the world, and are selected for activity-guided chromatographic purification work as a result of their potency and selectivity in panels of *in vitro* bioassays. The rationale for collecting the plants is somewhat different between these two projects, with species obtained by a combination of random screening and taxonomic selection for the antitumor agent project (*10,11*), while plants known to be edible are given priority in the cancer chemoprevention project (*12,13*). In both projects, however, it has been found that certain of the plants selected for fractionation have ethnomedical and/or agrochemical uses *per se*, or else they belong to a genus where for related species this is the case. It is hoped that the examples of bioactive compounds presented in the remainder of this chapter will be of interest to those whose primary interest is the discovery of novel natural products with potential agricultural applications.

Natural Products with Potential Antitumor Activity

In the past, antimicrobial- and plant-derived natural products have provided a substantial number of clinically used cancer chemotherapeutic agents, both in their native form and as lead compounds for synthetic optimization. Several such compounds, in addition to compounds from marine animals are under development as anticancer agents (14-17). Considerable support for future natural product anticancer drug discovery programs was obtained from a recent statistical survey, in which it was concluded that the already known natural products in selected databases show fundamental differences from existing libraries of synthetic compounds, in possessing higher molecular weights, and containing increased amounts of oxygen, with lower numbers of nitrogen, halogen, and sulfur atoms, and in representing a relatively larger proportion of compounds with sp^3-functionalized bridgehead atoms (18). Furthermore, it was concluded in this survey that there are still new types of natural products to discover (18).

Our present work directed to the discovery of novel cancer chemotherapeutic agents from plants is funded by the U.S. National Cancer Institute through the National Cooperative Natural Products Drug Discovery Group mechanism, and represents a collaboration between personnel at the College of Pharmacy, University of Illinois at Chicago, the Chemistry and Life Sciences Division, Research Triangle Institute in North Carolina, and Bristol-Myers Squibb, Pharmaceutical Research Institute, Princeton, NJ (10,11). Formerly, the industrial partner in our consortium was Glaxo Wellcome Medicines Research Centre, Stevenage, U.K. (10). Further to developing a formal agreement with official representatives in each country where plant acquisition is to be conducted, up to 500 primary plant samples are collected each year for this project, mainly from tropical rainforest areas. Crude organic-soluble extracts are made in a standardized manner (19), and these are then tested in a battery of *in vitro* bioassays (cell-based, enzyme inhibition, receptor-binding) housed at the various project sites. Plants are collected initially in sufficient quantities (*ca.* 0.5-1 kg dry weight) to perform activity-guided fractionation if the preliminary biological test data are supportive of this. Selected plant crude extracts and pure active compounds are subjected to evaluation in various secondary *in vitro* bioassays, inclusive of mechanism-of-action determinations, and promising compounds are tested in *in vivo* murine xenograft systems. Compounds of exceptional promise are considered for pharmaceutical development. As a result of the work carried out in this project to date, over 200 compounds active in one or more bioassays have resulted (10,11). Examples of some of these are presented in the remaining paragraphs of this section.

There is now substantial literature information on the insecticidal activity of constituents of species in the genus *Aglaia* (Meliaceae) (e.g., 20-22). Particularly active in this regard are the cyclopenta[b]benzofurans, for which Greger and co-workers have suggested the term "flavaglines" (22). Some of these compounds are very potent insecticides, and for example, several nitrogen-containing cyclopenta[b]benzofurans from the fruits of *A. elliptica* Bl. collected in Indonesia were active against the polyphagous insect pest *Spodoptera littoralis* at the 1 ppm level when incorporated in the diet (*21*). This plant appears to have had only limited use medicinally, with the bark utilized to treat tumors and leaves applied to wounds in the Philippines (*23*). In our own work on the stems and fruits of *A. elliptica* collected in Thailand, chloroform-soluble extracts of these plant parts were found to exhibit significant cytotoxic activity when evaluated against a small panel of human cancer cell lines (*23*). Activity-guided fractionation of the stems of *A. elliptica* led to the isolation of the known compound methyl rocaglate (**1**) and the novel cyclopenta[*b*]benzofurans **2**, **4**, and **5**, while similar treatment of the fruits led to the isolation of a further novel analog, **3**, in addition to **1** and **2** (*23*). When these five compounds were tested for cytotoxicity individually, compounds **1**, **4**, and **5** were the most potent, with **3** and **4** being markedly less active. For example, the most naturally abundant compound (**2**) obtained in our investigation exhibited ED_{50} values of 0.0009 and 0.0008 µg/mL, when evaluated against breast cancer (BC1) and glioblastoma (U373) cell lines (*23*). In a follow-up mechanism-of-action study on this compound, it was found that it acts as a cytostatic agent in cultured human lung cancer (Lu1) cells, and markedly reduced cellular protein synthesis, with much less effect on nucleic acid synthesis (*24*). Some evidence was obtained for the *in vivo* efficacy of compound **2** in athymic nude mice implanted subcutaneously with human breast cancer cells (*24*), and this substance is now being subjected to additional biological evaluation at the National Cancer Institute. Unlike the Indonesian collection of *A. elliptica* mentioned previously (*21*), no nitrogen-containing cyclopentabenzo[*b*]furans were isolated in our investigation on this same species collected in Thailand, which suggests the existence of different chemical races for this species (*23*).

	R_1	R_2	R_3
2	OH	H	COOMe
3	OH	H	H
4	=O		H
5	OCHO	H	COOMe

The insecticidal and insect antifeedant activities of *Azadirachta* species (Meliaceae) are well established, particularly *A. indica* A. Juss. (the Indian neem tree), which is the source of the commercially available insecticidal limonoid, azadiractin (*25,26*). Our group embarked on the study of a member of the genus *Azadirachta* which had not been studied phytochemically or biologically previously, namely, *A. excelsa* (Jack) Jacobs (*27*). The plant is documented as a timber source, with the older leaves used medicinally in Borneo (*28*). It was found that a chloroform-soluble extract of the stems of this species (collected in Thailand) was significantly cytotoxic against the KB (human epidermoid carcinoma of the mouth) cell line, and so this lead was selected for further investigation (*27*). However, since it was already known that the meliacin-type limonoids nimbolide (**6**) and 28-deoxonimbolide (**7**) are cytotoxic towards human cancer cells (*29*), these two known compounds were rapidly detected using a preliminary LC-ESMS/bioassay dereplication procedure (*27,30*). Two novel limonoids were isolated and characterized in our investigation on *A. excelsa*, namely, 2,3-dihydronimbolide (**8**) and 3-deoxynimbolide (**9**). On subsequent bioassay against a panel of human cancer cell lines, pure compounds **6** and **7** were moderately cytotoxic, with the most susceptible cell line found in each case to be LNCaP hormone-dependent human prostate cancer cells (ED$_{50}$ 0.9 and 1.9 µg/mL, respectively), while **8** and **9** were inactive (ED$_{50}$ >20 µg/mL) (*27*). It would be of interest to determine if compounds **8** and **9** have any activity against agrochemical pest targets.

	R$_1$	R$_2$	Other
6	O	O	$\triangle^{2,3}$
7	O	H$_2$	$\triangle^{2,3}$
8	O	O	----
9	α-OH	H$_2$	----

The small genus *Ratibida* is a member of the North American subtribe Rudbeckiinae (tribe Heliantheae, family Asteraceae), and its phytochemical profile is characterized by the elaboration of various types of sesquiterpene lactones (*31-33*). Two sesquiterpene lactones from *R. mexicana* (Wats.) Sharp, isoalloalantolactone and elema-1,3,11-trien-8,12-olide, were shown to inhibit the growth of three phytopathogenic fungi (*Fusarium oxysporum*, *Phythium* sp., *Helminthosporium* sp.), and to inhibit the radicle growth of *Amarathus hypochondiacus* and *Echinocloa crus-galli* (*33*). In our work, a different species of this genus was studied, namely, *R. columnifera* (Nutt.) Wood & Standl., which was previously used by Native Americans to treat fevers, chest pain, snakebite, and as an emetic (*28*). However, there was no prior information on the biological effects or phytochemical constituents of *R. columnifera*. Separate work-up of the flowers and leaves of this species, collected in Texas, led to the isolation of ten cytotoxic substances, using the LNCaP (hormone-dependent human prostate) cell line as a monitor (*34*). The majority of these compounds were broadly cytotoxic xanthanolide sesquiterpene lactones, including the known compound 9-oxo-*seco*-ratiferolide-5α-*O*-(2-methylbutyrate) (**10**), and the novel analogs **11-13**.

Of the various isolates obtained from this plant acquisition, compound **10** was the only one selected for follow-up biological and mechanistic studies. Thus, when evaluated in a 25-cell line tumor panel, it exhibited a mean IC_{50} value of 1.46 μM, and demonstrated a novel sensitivity pattern (*34*). This compound was investigated for its effects on the cell cycle and on apoptosis, and found to induce G_1 arrest in wild-type p53 A2780S ovarian cancer cells, while only G_2/M arrest was affected in p53 mutant A2780R ovarian cancer cells. Both these cell lines were found to undergo apoptosis as the result of treatment with compound **10**. At the concentration range 10-100 μM, compound **10** had no effects on tubulin polymerization, or on the catalytic ability of topoisomerase I and II enzymes, or on DNA intercalation, and it was concluded that this sesquiterpene acts as a cytotoxic compound via a novel mechanism of action (*34*). Unfortunately, compound **10** was not active in two *in vivo* xenograft evaluations carried out, using murine lung carcinoma (M109) and human colon carcinoma (HCT116) models, at the doses utilized (*34*).

	R_1	R_2
11	—o—	
12	H	H
13	H	OH

Natural Products with Potential Cancer Chemopreventive Activity

A second approach to intervening in the carcinogenesis phenomenon through the use of plant natural products refers to cancer chemoprevention. This term may be defined as "the prevention of cancer by the administration of one or more chemical entities, either as individual drugs or as naturally occurring constituents of the diet" (*35*). The goal of cancer chemoprevention is to block the carcinogenesis initiation and to arrest or reverse the progression of premalignant cells (*36*). There is now substantial evidence that phytochemical constituents of the diet in common beverages, culinary herbs, fruits, spices, and vegetables inhibit the development of cancer in standard long-term animal models (*37,38*). Natural compounds such as diallyl sulfide, ellagic acid, as well as isothiocyanates from *Brassica* species, exhibit carcinogenesis blocking (anti-initiating) effects, while curcumin, epigallocatechin gallate, and quercetin demonstrate suppressing (antipromotion/antiprogression) effects (*35*).

In our program to eludicate new cancer chemopreventive agents, funding by the U.S. National Cancer Institute has been provided through the Program Project mechanism, with all of the technical work conducted at the University of Illinois at Chicago, embracing plant acquisition, isolation chemistry, *in vitro* and *in vivo* biological evaluation, scale-up synthetic chemistry, and analytical and statistical support (*12,13*). The plants collected are both food plants and species collected in the field, and each acquisition is inventoried, dried, and milled, with small amounts of the comminuted material extracted with methanol and then partitioned into an organic solvent. Each dried organic-soluble residue is then assayed against a panel of about ten *in vitro* bioassays, with one of these used for activity-guided fractionation of active leads (*39*). Selected crude extracts and pure compounds are evaluated in a mouse mammary gland secondary assay, in which the ability to inhibit preneoplastic lesions induced by the carcinogen DMBA is determined (*39*). Very promising compounds may then be subjected to full-term carcinogenesis studies in animal models (*12,13,39*). In this project to date, well over 100 active natural products and their synthetic analogs have been obtained, and some of these have been subjected to substantial biological evaluation (*12,13,39*). Several examples are provided in the paragraphs below.

Mundulea sericea (Willd.) A. Chev. (syn. *M. suberosa* Benth.) (Leguminosae) is a shrub occurring in central and southern Africa, and in India, and its various plant parts have been used for aphrodisiac, insecticidal, and piscidicidal purposes, and to treat diarrhea and stomachache (*28,40,41*). During our screening program for potential cancer chemopreventive agents, a sample of the bark of *M. sericea* collected in Kenya was chosen for further work, because its ethyl acetate-soluble extract was potently active in a cell culture phorbol ester-mediated ornithine decarboxylase (ODC) inhibition assay (*39-41*). Following activity-guided fractionation using this assay, a number of active flavonoids were obtained, inclusive of the two known rotenoids deguelin (**14**) and tephrosin (**15**), and their novel analogs, (-)-13α-hydroxydeguelin (**16**) and 13α-hydroxytephrosin (**17**) (*40*). Deguelin (**14**) is a major constituent of cubé resin insecticide (obtained from the roots of *Lonchocarpus utilis* A.C. Smith and *L. urucu* Killip and Smith) (*42*). However, in our work, this compound was the most potent of the *M. sericea* constituents in the ODC inhibition assay, showing an IC_{50} value of 0.3 ng/mL (*40*). In contrast tephrosin (**15**) was over two orders of magnitude less active than deguelin (**14**), and the novel hydroxylated derivative **16** was about 10 times less active than the parent compound (*40*). Accordingly, deguelin (**14**) was subjected to further biological evaluation, and was found to be active in the mouse mammary organ culture assay, and in two long-term animal models (a skin carcinogenesis model with CD-1 mice and a mammary carcinogenesis model with Sprague-Dawley rats) (*43*). In addition, deguelin (**14**) was evaluated mechanistically (*44*). Owing to its extreme potency and effectiveness at inhibiting cancer in the animal models, deguelin (**14**) is one of the best leads to date to emerge from our natural product cancer chemoprevention program (*13*). The isoflavonoid, munetone (**18**), another of the ODC-inhibitory flavonoids to be obtained from our investigation on *M. sericea* bark, was found to inhibit DMBA-induced preneoplastic lesions in the mouse mammary organ culture system (*45*).

Tephrosia purpurea Pers. (Leguminosae) is pantropical coastal shrub, with a wide variety of ethnobotanical uses. Various parts of the plant employed in India

	R_1	R_2
14	H	H
15	OH	H
16	H	OH
17	OH	OH

for the treatment of asthma, diarrhea, gonorrhea, and rheumatism (*46*), and the seeds are used for their insecticidal and insect repellant properties (*47*). While the effect on insects of this plant are attributed to the presence of various rotenoids (*28*), this compound class was not found in our own work on this species. When the flowering and fruiting parts of *T. purpurea* were investigated, non-polar extracts were found to significantly induce quinone reductase (QR) in Hepa 1c1c7 hepatoma cells (*46*). Activity-guided fractionation using this bioassay led to the isolation of several active flavonoids, including the novel isoflavone, 7,4'-dihydroxy-3',5'-dimethoxyisoflavone (**19**) and the novel chalcone (+)-tephropurpurin (**20**). The latter compound was obtained in very small amounts (0.00005% w/w yield), but was the most interesting compound of the active isolates from *T. purpurea*, because it exhibited a CD (concentration to double QR activity) value of 0.15 μM, while it was not particularly cytotoxic for the cultured cells used in the assay (*46*). Since our initial publication of the structure of (+)-tephropurpurin (*46*), the relative stereochemistry at C-2'', C-3'' and C-4'' has been revised so as to be consistent with a recent X-ray crystallographic study published on the related compound, (+)-purpurin (*48*). Additional work has been carried out on this lead, and three novel flavonoids, (+)-tephrorin A (**21**) and (+)-tephrosone (**22**), and **21** and **22** were found to have

CD values in the QR assay of 4.0 and 3.1 μM, respectively. The absolute configuration for both compounds **21** and **22** was determined by Mosher ester methodology (*49*).

Summary and Conclusions

For the five medicinal plants chosen to exemplify recent phytochemical progress in our two current projects directed towards the discovery of novel plant-derived antitumor agents and natural product cancer chemopreventive agents, respectively, several previously known and new secondary metabolites representing variety of structural types have been obtained. The structural classes of the bioactive substances described in this chapter comprise chalcones,

cyclopenta[*b*]benzofurans, a flavanone, isoflavones, limonoids, rotenoids, and sesquiterpene lactones. All five plants selected represent species or genera for which there is information available on pesticidal activity. Further evidence is provided herein for the biological cross-reactivity of plant secondary metabolites in different test systems.

Acknowledgments

The authors of this chapter gratefully acknowledge two grant awards from the National Cancer Institute, NIH, Bethesda, MD, which have led to the isolation and structure elucidation of the compounds described in this chapter, namely, CA52956 (potential anticancer agents; P.I., A.D. Kinghorn) and CA48112 (potential cancer chemopreventives, P.I., J.M. Pezzuto). We also wish to thank many outstanding colleagues and postdoctoral and graduate student associates who have participated in these two projects, and whose names are indicated in the bibliography below.

References

1. *Biologically Active Natural Products. Potential Use in Agriculture;* Cutler, H. G., Ed.; Symposium Series No. 380; American Chemical Society: Washington, D.C., 1988; 483 pp.
2. *Human Medicinal Agents from Plants;* Kinghorn, A. D.; Balandrin, M. F., Eds.; Symposium Series No. 534; American Chemical Society: Washington, D.C., 1993; 356 pp.
3. Buta, J. G.; Kalinski, A. In *Biologically Active Natural Products. Potential Use in Agriculture;* Cutler, H.G., Ed.; Symposium Series No. 380; American Chemical Society: Washington, D.C., 1988, pp 294-304.
4. Duke, S. O.; Paul, R. N., Jr.; Lee, S. M. In *Biologically Active Natural Products. Potential Use in Agriculture;* Cutler, H.G., Ed.; Symposium Series No. 380; American Chemical Society; Washington, D.C.; 1988; pp 318-334.
5. Chew, F. S. In *Biologically Active Natural Products. Potential Use in Agriculture;* Cutler, H.G., Ed.; Symposium Series No. 380; American Chemical Society: Washington, D.C., 1988; pp 155-181.
6. Betz, J. M.; Fox, W. D. In *Food Phytochemicals for Cancer Prevention I. Fruits and Vegetables;* Huang, M.-T.; Osawa, T.; Ho, C.-T.; Rosen, R. T., Eds.; Symposium Series No. 546, 1994; pp 181-196.
7. Balandrin, M. F.; Klocke, J. A.; Wurtele, E. S.; Bollinger, W. H. *Science* **1985**, *228*, 1154-1160.
8. Williams, D. H.; Stone, M. J.; Hauck, P. R.; Rahman, S. K. *J. Nat. Prod.* **1989**, *52*, 1189-1208.
9. Wedge, D. E.; Camper, N. D. In: *Biologically Active Natural Products: Pharmaceuticals;* Cutler, S. J.; Cutler, H. G., Eds.; CRC Press: Boca Raton, FL, 1999, pp 1-15.
10. Kinghorn, A. D.; Farnsworth, N. R.; Beecher, C. W. W.; Soejarto, D. D.; Cordell, G. A.; Pezzuto, J. M.; Wall, M. E., Wani, M. C.; Brown, D. M.; O'Neill, M. J.; Lewis, J. A.; Besterman, J. M. In *New Trends in Natural Products Chemistry;* Atta-ur-Rahman; Choudhary, M. I.; Eds.; Harwood Academic Publishers: Amsterdam, The Netherlands; 1998; pp 79-94.
11. Kinghorn, A. D.; Farnsworth, N. R.; Soejarto, D. D.; Cordell, G. A.; Pezzuto, J. M.; Udeani, G. O.; Wani, M. C.; Wall, M. E.; Navarro, H. A.;

112

Kramer, R. A.; Menendez, A. T.; Fairchild, C. R.; Lane, K. E.; Forenza, S.; Vyas, D. M.; Lam, K. S.; Shu, Y.-Z. *Pure Appl. Chem.*, in press.

12. Pezzuto, J. M.; Beecher, C. W. W.; Fong, H. H. S.; Farnsworth, N. R.; Mehta, R. G.; Moon, R. C.; Hedeyat, S.; Udeani, G. O.; Moriarty, R. M.; Kinghorn, A. D. In *New Trends in Natural Product Chemistry;* Atta-ur-Rahman; Choudhary, M. I., Eds.; Harwood Academic Press: Amsterdam, The Netherlands; 1998, pp 95-107.

13. Kinghorn, A. D.; Fong, H. H. S.; Farnsworth, N. R.; Mehta, R. G.; Moon, R. C.; Moriarty, R. M.; Pezzuto, J. M. *Curr. Org. Chem.* **1998**, *2*, 597-612.

14. Chabner, B. A.; Allegra, C. J.; Curt, C. A.; Calabresi, P. In *Goodman & Gilman's The Pharmacological Basis of Therapeutics;* Hardman, J. G.; Limbird, L. E.; Molinoff, R. W.; Ruddon, R. W.; Gilman, A. G., Eds.; McGraw-Hill: New York; 1996; 9th Edn.; pp 1233-1287.

15. Cragg, G. M.; Newman, D. J.; Snader, K. M. *J. Nat. Prod.* **1997**, *61*, 52-60.

16. Shu, Y.-Z. *J. Nat. Prod.* **1998**, *61*, 1053-1071.

17. Cragg, G. M.; Newman, D. J. *Cancer Invest.* **1999**, *17*, 153-163.

18. Henkel, T.; Brunne, R. M.; Müller, H.; Reichel, F. *Angew. Chem. Int. Ed.* **1999**, *38*, 643-647.

19. Wall, M. E.; Wani, M. C.; Brown, D. M.; Fullas, F.; Oswald, J. B.; Josephson, F. F.; Thornton, N. M.; Pezzuto, J. M.; Beecher, C. W. W.; Farnsworth, N. R.; Kinghorn, A. D. *Phytomedicine* **1996**, *3*, 281-286.

20. Ishibashi, F.; Satasook, C.; Isman, M. B.; Towers, G.H.N. *Phytochemistry* **1993**, *32*, 307-310.

21. Nugroho, B. W.; Güssregen, B.; Wray, V.; Witte, L; Bringmann, G.; Proksch, P. *Phytochemistry* **1997**, *45*, 1579-1585.

22. Brader, G.; Vajrodaya, S.; Greger, H.; Bacher, M.; Kalchhauser, H.; Hofer, O. *J. Nat. Prod.* **1998**, *61*, 1482-1490.

23. Cui, B.; Chai, H.; Santisuk, T.; Reutrakul, V.; Farnsworth, N. R.; Cordell, G. A.; Pezzuto, J. M.; Kinghorn, A. D. *Tetrahedron* **1997**, *53*, 17625-17632.

24. Lee, S. K.; Cui, B.; Mehta, R. R.; Kinghorn, A. D.; Pezzuto, J. M. *Chem.-Biol. Interact.* **1998**, *115*, 215-228.

25. Lee, S. M.; Klocke, J. A.; Barnby, M. A.; Yamasaki, R. B.; Balandrin, M. F. In *Naturally Occurring Pest Bioregulators*; Heydin, P. A., Ed.; Symposium Series No. 449; American Chemical Society: Washington, D.C.; 1991; pp 293-304.

26. Harborne, J. B. In *Ecological Chemistry and Biochemistry of Plant Terpenoids. Proceedings of the Phytochemical Society of Europe, vol. 31*; Harborne, J. B.; Tomas-Barbaran, F. A., Eds.; Oxford, U.K.; Oxford University Press, 1991; pp 399-426.

27. Cui, B.; Chai, H.; Constant, H. L.; Santisuk, T.; Reutrakul, V.; Beecher, C. W. W.; Farnsworth, N. R.; Cordell, G. A.; Pezzuto, J. M.; Kinghorn, A. D. *Phytochemistry* **1998**, *47*, 1283-1287.

28. Hocking, G. M. *A Dictionary of Natural Products;* Plexus Publishing Co.: Medford, NJ, 1997.
29. Kigodi, P. G. K.; Blaskó, G.; Thebtaranonth, Y.; Pezzuto, J. M.; Cordell, G. A. *J. Nat. Prod.* **1989**, *52*, 1246-1251.
30. Constant, H. L.; Beecher, C. W. W. *Nat. Prod. Lett.* **1995**, *6*, 193-196.
31. Ellmauerer, E.; Pathak, V. P.; Jakupovic, J.; Bohlmann, F.; Dominguez, X. A.; King, R. M.; Robinson, H. *Phytochemistry* **1987**, *26*, 159-163.
32. Jimenez, A.; Pereda-Miranda, R.; Bye, R.; Linares, E.; Mata, R. *Phytochemistry* **1993**, *34*, 1079-1082.
33. Calera, M. R.; Soto, F.; Sanchez, P.; Bye, R.; Hernandez-Bautista, B.; Anaya, A. L.; Lotina-Hennsen, B.; Mata, R. *Phytochemistry* **1995**, *40*, 419-425.
34. Cui, B.; Lee, Y. H.; Chai, H.; Tucker, J. C.; Fairchild, C. R.; Raventos-Saurez, C.; Long, B.; Lane, K. E.; Menendez, A. T.; Beecher, C. W. W.; Cordell, G. A.; Pezzuto, J. M.; Kinghorn, A. D. *J. Nat. Prod.* **1999**, *62*, 1545-1550.
35. Morse, M. A.; Stoner, G. D. *Carcinogenesis* **1993**, *14*, 1737-1746.
36. Sporn, M. B. *Lancet* **1993**, *342*, 1211-1212.
37. *Food Phytochemicals for Cancer Prevention I. Fruits and Vegetables;* Huang, M.-T.; Osawa, T.; Ho, C.-T.; Rosen, R. T., Eds.; Symposium Series No. 546; American Chemical Society: Washington, D.C., 1994; pp 427.
38. *Food Phytochemicals for Cancer Prevention II. Teas, Spices, and Herbs;* Ho, C.-T.; Osawa, T.; Huang, M.-T.; Rosen, R. T., Eds.; Symposium Series No. 547; American Chemical Society: Washington, D.C., 1994; pp 370.
39. Pezzuto, J. M.; Song, L. L.; Lee, S. K.; Shamon, S.; Mata-Greenwood, E.; Jang, M.; Jeong, H-J.; Pisha, E.; Mehta, R. G.; Kinghorn, A. D. In *Chemistry, Biological and Pharmacological Properties of Medicinal Plants from the Americas;* Hostettmann, K.; Gupta, M. P.; Marston, A., Eds.; Harwood Academic Publishers: Amsterdam, The Netherlands, 1998, pp 81-110.
40. Luyengi, L.; Lee, I.-S.; Mar, W.; Fong, H. H. S.; Pezzuto, J. M.; Kinghorn, A. D. *Phytochemistry* **1994**, *36*, 1523-1526.
41. Mbwambo, Z. H.; Luyengi, L.; Kinghorn, A. D. *Int. J. Pharmacog.* **1996**, *34*, 335-343.
42. Fang, N.; Casida, J. E. *J. Nat. Prod.* **1999**, *62*, 205-210.
43. Udeani, G. O.; Gerhäuser, C.; Thomas, C. F.; Moon, R. C.; Kosmeder, J. W.; Kinghorn, A. D.; Moriarty, R. M.; Pezzuto, J. M. *Cancer Res.* **1997**, *57*, 3424-3428.
44. Gerhäuser, C.; Lee, S. K.; Kosmeder, J. W.; Moriarty, R. M.; Hamel, E.; Mehta, R. G.; Moon, R. C.; Pezzuto, J. M. *Cancer Res.* **1997**, *57*, 3429-3435.

45. Lee, S. K.; Luyengi, L.; Mar, W.; Gerhäuser, C.; Lee, K.; Mehta, R. G.; Kinghorn, A. D.; Pezzuto, J. M. *Cancer Lett.* **1999**, *136*, 59-65.
46. Chang, L. C.; Gerhäuser, C.; Song, L.; Farnsworth, N. R.; Pezzuto, J. M., Kinghorn, A. D. *J. Nat. Prod.* **1999**, *60*, 869-873.
47. Saxena, B. N.; Dubey, D. N.; Nair, A. L. *Def. Sci. J.* **1974**, *24*, 43-48.
48. Pirrung, M. C.; Lee, Y. R.; Morehead, A. T.; McPhail, Jr, A. T. *J. Nat. Prod.* **1998**, *61*, 89-91.
49. Chang, L. C.; Chávez, D.; Song, L. L.; Farnsworth, N. R.; Pezzuto, J. M.; Kinghorn, A. D. *Org. Lett.* **2000**, *2*, 515-518.

Agricultural Biotechnology

Traditional breeding and agricultural chemistry have impressively increased the yields of food and feed worldwide during the past three decades. However, in order to meet the needs of an expanding world population new technologies that sustain the environment, and provide more nutritious and safe food are necessary. An understanding of the processes by which plants biotransform natural and synthetic materials and protect themselves against pests together with the tools of biotechnology have the potential to significantly increase the number of crop varieties grown while reducing production costs, protecting the land and providing a nutritious and safe food supply. The success of these new technologies will require the establishment of food, feed and environmental safety through the conduct of extensive studies which are independently reviewed and approved around the world.

The following chapters give a glimpse into the promise of agricultural biotechnology, and the challenges that face these new approaches. Chapters by H. Ohkawa and R. Edwards describe the development of herbicide resistant plants by the expression of specific cytochrome P450 monooxygenases and glutathione S-transferases, respectively. K. Lawton discusses the use of chemical activators of systemic acquired resistance (SAR) and T. Arie describes approaches for the identification of natural plant defenses for disease resistance. In the final chapter, J. Astwood summarizes the steps necessary to assure the food, feed and environmental safety of genetically modified crops.

Hideo Ohkawa, Ph.D.
Kobe University
Rokkodai-cho 1-1, Nada-Ku
Kobe 657-8501
JAPAN

William P. Ridley, Ph.D.
Monsanto Company
700 Chesterfield Pky. N.
St. Louis, MO 63198

Chapter 10

Herbicide Resistant Transgenic Plants Expressing Cytochrome P450 Monooxygenases Metabolizing Xenobiotics

Hideo Ohkawa[1], Hideyuki Inui[1], Yoshiro Imajuku[1], Hiromasa Imaishi[1], and Yasunobu Ohkawa[2]

[1]Department of Biological and Environmental Science, Faculty of Agriculture, Kobe University, Rokkodai-cho 1–1, Nada-ku, Kobe 657–8501, Japan
[2]National Institute of Agrobiological Resources, Ministry of Agriculture, Forestry and Fisheries, Tsukuba 305–8602, Japan

Cytochrome P450 (P450 or CYP) monooxygenases play an important role in the Phase I reactions of biotransformations of xenobiotics in higher plants. Some P450 species were selected for herbicide metabolism and then expressed in potato and rice plants. Human CYP1A1, CYP2B6 and CYP2C19 co-expressed in potato plants coordinately functioned, actively metabolized a number of herbicides and conferred cross-resistance to the corresponding herbicides. The transgenic rice plants expressing human CYP2C9 metabolized sulfonylurea herbicides and exhibited cross-resistance to these herbicides. The transgenic plants expressing P450 species metabolizing xenobiotics appear to be useful for phytoremediation of environmental contaminants.

Xenobiotics including herbicides are transformed in higher plants. The biotransformation processes are grouped into three main phases known as Phase I or conversion, Phase II or conjugation, and Phase III or compartmentation. Phase I reactions include oxidations, reductions and hydrolyses, whereas Phase II reactions are conjugations with glutathione, sugars or amino acids. In Phase

III, xenobiotic conjugates are converted to secondary conjugates or insoluble bound residues and are deposited in the vacuole or other compartments of plant cells (1,2).

The Phase I reactions are mostly catalyzed by cytochrome P450 monooxygenases and esterases. Cytochrome P450 monooxygenases localized on the microsomes consist of a number of cytochrome P450 (P450 or CYP) species and a generic NADPH-cytochrome P450 oxidoreductase (P450 reductase), catalyzing oxidative reactions of endogenous and exogenous lipophilic compounds including herbicides. Currently, over 400 P450 sequences from plants are known, but the function for about only 30 species have been identified. Important P450 enzymes whose characterization remains elusive or poorly understood are involved in the biosynthesis of sterols, fatty acids, glucosinolates, phenylpropanoids/flavonoids, salicylic acid, jasmonic acid, gibberellic acids, brassinosteroids and alkaloids, and in the metabolism of xenobiotics including herbicides (3,4). The Phase II of xenobiotic metabolism was mediated by glutathione S-transferases (GST). GST isozymes of maize in xenobiotic detoxification are the most studied and GST IV was found to play a major role in the tolerance of maize to the herbicide alachlor (5). The role of glucosyltransferases associated with xenobiotic metabolism is also well studied. Herbicide glutathionation and glucosylation take place in the cytosol. Both glutathione and glucose conjugates are transported into the vacuole by ATP-stimulated transporters in Phase III (6).

Plant P450 Species Metabolizing Herbicides

P450 enzymes metabolize herbicides such as atrazine, bentazon, chlorimuron, chlorsulfuron, chlortoluron, diclofop, fluometuron linuron, metolachlor, monuron, primisulfuron and triasulfuron (4). However, molecular information on these P450 species was limited, since it was rather difficult to identify a function of a P450 species in a large gene family. In addition, the activity of plant enzymes towards herbicides was low and unstable as compared with those of mammalian ones.

It was found that tobacco cultured S401 cells treated with 2,4-D metabolized chlortoluron to give ring-methyl hydroxylated and N-demethylated metabolites, whereas the cells hardly produced these metabolites without 2,4-D treatment. Based on these results, we attempted to clone cDNAs for P450 species metabolizing chlortoluron in tobacco cultured S401 cells treated with 2,4-D. As a result, four novel P450 cDNA clones were isolated. Based on their sequences, these were named as CYP71A11, CYP81B2, CYP81C1 and CYP81C2. Northern blot analysis with the cloned cDNAs revealed that both CYP71A11 and CYP81B2 were highly induced in the S401 cells treated with 2,4-D. Therefore, we sought to express each of both cDNA clones in the yeast *Saccharomyces cerevisiae*. Both CYP71A11 and CYP81B2 expressed in the yeast together with yeast P450 reductase or tobacco P450 reductase showed a higher 7-ethoxycoumarin O-deethylase activity than that of the control yeast cells. In addition, the yeast cells expressing both CYP71A11 and yeast P450 reductase exhibited enhanced N-demethylation towards chlortoluron, whereas the yeast cells expressing both CYP81B2 and tobacco P450 reductase showed a slightly enhanced ring-methyl hydroxylation. Therefore, both CYP71A11 and CYP81B2 were found to be involved in the metabolism of chlortoluron in the S401 cells treated with 2,4-D (7).

Feldmann et al. (8), described a reverse genetics approach making use of the large collection of T-DNA (the region of the Ti plasmid responsible for tumor formation) insertion mutants in *Arabidopsis thaliana* and the vast array of DNA sequences to identify insertion mutants for P450 genes. The isolated mutants are then screened for specific alteration in plant growth and development as related to P450 species. A substantial database of partially sequenced cDNA clones or ESTs (Expressed Sequence Tags) has been increasing due to the efforts of a number of laboratories worldwide. Combined with these many attributes is undoubtedly an ever growing population of insertion mutants. We found that CYP86A1 isolated from *A. thaliana* metabolized chlortoluron as well as fatty acids (9). This approach appears to be useful for characterization of physiological function of novel P450 species in plants.

It was also reported that CYP71A10, CYP71B1, CYP73A1, CYP76B1 and CYP81B1 from plants also metabolized chlortoluron, although CYP81B1 and CYP86A1 metabolized endogenous fatty acids as well, as shown in Table 1 (4). In addition, CYP71C3v2 metabolize triasulfuron. It is important to identify P450 species metabolizing herbicides in monocotyledonous plants including rice plants. Particularly, these P450 species metabolizing herbicides seem to be important targets for herbicide safeners, which induced herbicide metabolism enzymes such as GST and P450.

Table 1. Plant P450 Species Metabolizing Herbicides

P450 species	Substrate	
	Endogenous	*Exogenous*
CYP71A10	-	phenylurea herbicides
CYP71A11	-	chlortoluron
CYP71C3v2	-	triasulfuron
CYP73A1	-	chlortoluron
CYP76B1	-	phenylurea herbicides
CYP81B1	Fatty acids	chlortoluron
CYP81B2	-	chlortoluron
CYP86A1	Fatty acids	chlortoluron

Mammalian P450 Species Metabolizing Herbicides

The commercial herbicide chemicals have been examined for metabolism in mammals from the standpoint of toxicological evaluation. Microsomal P450 monooxygenases in mammalian livers are well known as drug-metabolizing enzymes involved in oxidative metabolism of xenobiotics including herbicides. However, it was not identified yet which P450 species catalyze oxidative reactions of a particular herbicide. It was reported that 11 P450 species in human livers covered over 90% of P450-dependent drug metabolism. These P450 species are CYP1A1, CYP1A2, CYP2A6, CYP2B6, CYP2C8, CYP2C9, CYP2C18, CYP2C19, CYP2D6, CYP2E1 and CYP3A4 (10). These P450 species are known to exhibit broad and overlapping substrate specificity towards xenobiotics. Therefore, we examined herbicide metabolism in the recombinant yeast strains expressing each of 11 human P450 species (3). Chlortoluron and atrazine were found to be metabolized through ring-methyl hydroxylation and N-demethylation by CYP1A1, CYP1A2, CYP2C19 and CYP2D6. These four P450 species catalyzed the same reaction towards both herbicides. It was also found that the specific activity of human CYP1A1 expressed in the yeast towards chlortoluron was higher as compared with those of tobacco CYP81B2 and CYP71A11. CYP2B6 and CYP2C19 also metabolized pyributicarb to give *m-t*-butylphenol. Sulfonylurea herbicides including chlorsulfuron and imazosulfuron were specifically metabolized by CYP2C9 to give the corresponding ring-hydroxylated metabolites. Among 11 human P450 species, four major species, CYP1A1, CYP2B6, CYP2C9 and CYP2C19 were found to metabolize 14, 11, 7 and 7 herbicide chemicals, respectively, among the chemicals examined, as listed in Table 2. Thus, it was found that each of P450

Table 2. Human P450 Species Metabolizing Herbicides

P450 species	Herbicide
CYP1A1	atrazine, benfuresate, chlortoluron, diuron, esprocarb, isoxaben, mefenacet, methabenzthiazuron, norflurazon, pyrazoxyfen, pyriminobac-methyl, quizalofop-ethyl, simazine and simetryn
CYP2B6	acetochlor, alachlor, benfuresate, chloridazon, esprocarb, ethofumesate, mefenacet, metolachlor quinclorac, simetryn and trifluralin
CYP2C9	benfuresate, chlorsulfuron, diuron, imazosulfuron, pyriminobac-methyl, triasulfuron and simetryn
CYP2C19	acetochlor, atrazine, benfuresate, chlortoluron, diuron, esprocarb, ethofumesate, mefenacet, methabenzthiazuron, metolachlor, norflurazon, pyributicarb, pyriminobac-methyl, simazine, simetryn, bensulfuron-methyl and chloridazon

____ ; suspected environmental endocrine disruptors

species metabolizing xenobiotics metabolized a number of herbicide chemicals with different chemical structures and modes of herbicide action. Among them, suspected environmental endocrine disruptors were included.

Genetic Engineering of Herbicide Tolerant Plants

Since each of the selected human P450 species actively metabolized a number of herbicide chemicals with different structures and modes of herbicide action, we attempted to express these P450 species in plants for enhancement of herbicide metabolism.

Potato is one of the major food crops facilitated with an efficient transformation method for gene engineering. We constructed four recombinant plasmids for expression of each of human CYP1A1, CYP2B6 and CYP2C19 as well as for co-expression of these three P450 species, as shown in Fig 1 (11), (12). Each of the constructed plasmids was subjected to transformation of microtuber discs of potato (*Solanum tuberosum cv. May Queen*) by the use of *Agrobacterium tumefaciens* strains LBA4404 and C58C1 containing each of the plasmids. Kanamycin resistant shoots were selected for PCR (polymerase chain reaction) to detect a transgene, northern blotting to detect a mRNA and western blotting to detect a P450 protein. Finally, the transgenic plants producing the highest amounts of P450 proteins were selected: S1965 (CYP1A1), S1972 (CYP2B6), S1974 (CYP2C19) and T1977 (CYP1A1, CYP2B6 and CYP2C19).

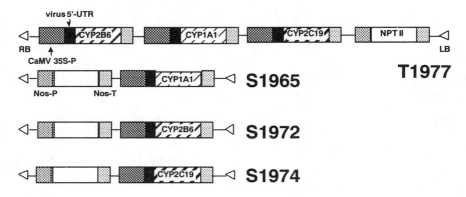

Figure 1. Expression plasmids and selected potato transformants

The expression levels of CYP1A1 and CYP2B6 were almost similar in the transgenic plants expressing each of the P450 species and co-expressing three

P450 species, whereas the level of CYP2C19 was higher in T1977 than in S1974 (CYP2C19). When ^{14}C-atrazine was added to a nutrient solution at 5μM, four transgenic plants and the control plant took up ^{14}C similarly. ^{14}C was extracted from the plants 8 days after treatment and then analyzed by thin-layer chromatography (TLC). Atrazine remained was higher in the control plant, but lesser in the transgenic plants. Particularly, atrazine remained was the lowest in T1977. In stead, the amount of the metabolite 6-chloro-2,4-diamino-1,3,5-triazine (DIDE) was higher in T1977 and S1965 and the lowest in the control. Therefore, CYP1A1 expressed in potato plants actively metabolized atrazine, probably by the interaction of CYP1A1 with endogenous P450 reductase.

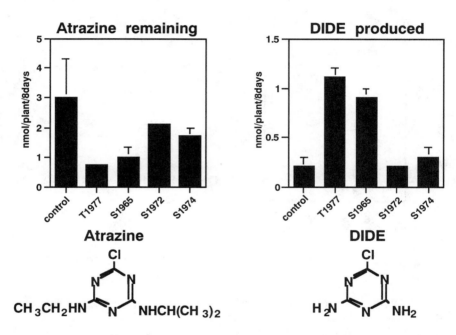

Figure 2. ^{14}C-Atrazine metabolism in the transgenic potato plants

^{14}C-Chlortoluron added to a nutrient solution at 20μM was taken up by both transgenic and control plants similarly. After 12 hours, ^{14}C extracted from the plants was analyzed by TLC. Chlortoluron remained was the highest in the control, lower in the transgenic plants and particularly lowest in T1977 and S1965. On the other hand, ring-methyl hydroxylated and ring-methyl hydroxylated N-demethylated metabolites were produced in T1977 and S1965 at higher levels, but produced in the control at the lowest level. Thus, CYP1A1 expressed in potato plants also actively metabolized chlortoluron (CT) (13).

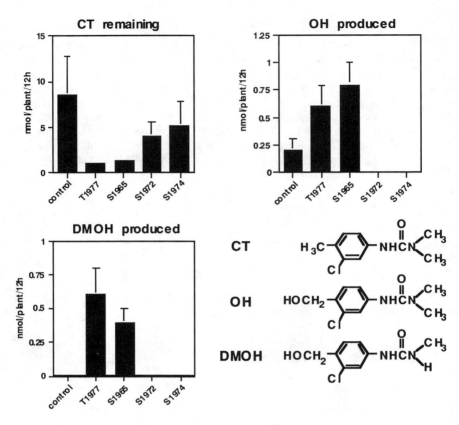

Figure 3. ^{14}C-Chlortoluron metabolism in the transgenic potato plants

^{14}C-Pyributicarb (PC) added to a nutrient solution at 10μM gave control at the highest level, and in T1977 at the lowest level 8 days after treatment. Instead, the metabolite *m-t*-butylphenol (BP) was produced at the highest level in T1977 and at the lowest in the control. Therefore, CYP2C19 expressed in potato plants coordinately functioned with the other two P450 species in T1977 and actively metabolized pyributicarb, whereas CYP2C19 expressed in S1974 was not so active in the metabolism of the herbicide. When the transgenic plants were assayed for herbicide resistance, T1977 showed higher resistance to atrazine, chlortoluron and pyributicarb as compared with the plants expressed individual P450 species. Probably, on co-expression of CYP1A1, CYP2B6 and CYP2C19 in the potato plants, these P450 species coordinately functioned in the metabolism of the herbicide chemicals to exhibit higher resistance to the corresponding herbicides. The produced metabolites mostly remained in the potato plants as their conjugates with glucose.

It was also found that transgenic potato plants co-expressing CYP1A1, CYP2B6 and CYP2C19 actively metabolized the insecticide methoxychlor through *O*-demethylation to produce mono-demethylated and di-demethylated metabolites, which were suspected environmental endocrine disruptors (14). However, these metabolites were deposited as glucose conjugates in the plants.

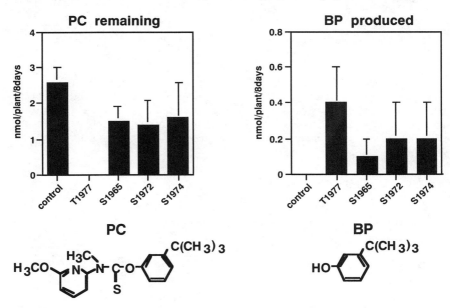

Figure 4. ^{14}*C-Pyributicarb metabolism in the transgenic potato plants*

Sulfonylurea herbicides were highly selective and showed a wide spectrum for weeds at a low dosage. However, they had potential problems with the wide usage on emergence of herbicide-tolerant weeds. In addition, rice plants were lower in metabolism and sensitive to chlorsulfuron and triasulfuron. It was found that the sulfonylurea herbicides were specifically metabolized by CYP2C9. Therefore, we attempted to express human CYP2C9 in rice plants (11). The recombinant plasmid for expression of CYP2C9 was constructed as shown in Fig 5. Calli of rice plants (*Oryza sativa L. cv. Nipponbare*) were transformed by the use of *A. tumefaciens* EHA101 containing the plasmid. Regenerated plants were selected on hygromycin. Rice seeds selected were further screened on chlorsulfuron. Finally, the selected transgenic plant 2C9-57-

E-11 was examined for southern blot, northern blot and western blot analyses to confirm the expression of the transgene.

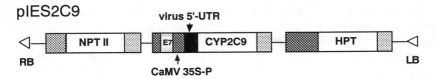

Figure 5. A expression plasmid for transformation of rice plants

^{14}C-Chlorsulfuron was added to a nutrient solution at 1.25μM. ^{14}C containing fractions were extracted from the plants and analyzed by TLC. Chlorsulfuron remained was higher in the control than in the transgenic one. Instead, more metabolites were produced in the transgenic plants as compared with the control. The major metabolites were glucosides of ring-hydroxylated chlorsulfuron.

Chlorsulfuron   **Hydroxylated chlorsulfuron**   **Glucose conjugate**

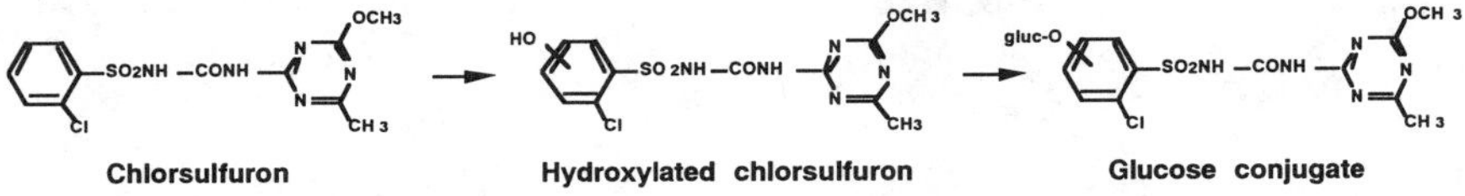

Figure 6. Proposed metabolic pathways for chlorsulfuron in the transgenic rice

^{14}C-Imazosulfuron added to a nutrient solution at 1.25μM and then ^{14}C containing fractions were extracted from the plants and analyzed by TLC after 36 hours. Imazosulfuron remaining was higher in the control than in the transgenic plants. On the other hand, more metabolites were produced in the transgenic plants as compared with the control. The major products were glucosides of hydroxylated imazosulfuron. When 50nM of chlorsulfuron and 5μM of imazosulfuron were each added to a MS (Murashige and Skoog) medium, the transgenic rice plants showed resistance to both herbicides. Thus, it was found that CYP2C9 expressed in the rice plants actively metabolized both sulfonylurea herbicides and exhibited cross-resistance to both herbicides.

Based on these results, it was found that expression of the human P450 species metabolizing herbicides in the plants actively enhanced metabolism of the herbicides and conferred resistance to the herbicides. The metabolites produced by the action of the P450 species mainly remained as glycosides of the herbicides in the plants. These transgenic plants expressing human P450 species appear to show potential for breeding crops with herbicide resistance as well as low pesticide residues. These plants are also important for phytoremediation of environmental contaminants, since P450 species show a wide substrate specificity.

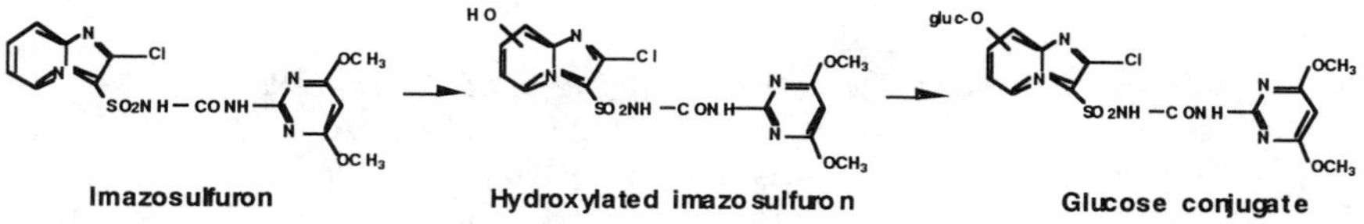

Figure 7. Proposed metabolic pathways for imazosulfuron in the transgenic rice

Concluding Remarks

Before practical use, the transgenic plants must be evaluated in the field for efficacy as well as risk assessment. The safety of the transgenes and their gene products, the properties of the modified plants, the expression stability of the transgene and the safety of the modified plants must be considered for risk assessment. Potential interactions of the modified plants with wild relatives and even the remote probability of undesired interaction with other organisms must be evaluated. In the case of P450 species metabolizing xenobiotics, these show broad and overlapping substrate specificity towards a wide variety of lipophilic chemicals with different structures including natural products. For example, CYP1A2 and CYP2A6 metabolized caffeine and nicotine, respectively. Therefore, the metabolism of xenobiotics as well as secondary metabolites produced in the plants must be considered. The introduced P450 species may metabolize phytoalexins. Therefore, the transgenic plants must be evaluated for change of sensitivity to pathogens as well as insect pests.

Acknowledgement

The Program for Promotion of Basic Research Activities for Innovative Bioscience of the Bio-oriented Technology Research Advancement Institution of Japan supported this study in part.

References

1. Kreuz, K.; Tommasini, R.; Martinoia, E.; *Plant Physiol* **1996**, *111*, 349-353.
2. *Regulation of Enzymatic Systems Detoxifying Xenobiotics in Plants;* Hatzios, K, K.; Hatzios, K, K., Ed.; Kluwer Academic Publishers, the Netherlands, **1977**; pp1-5.

3. Ohkawa, H.; Tsujii , H.; Shimoji, M.; Imajuku, Y.; Imaishi, H.; *J Pesticide Sci.* *1999, 24,* 197-203.

4. Ohkawa, H.; Imaishi, H.; Shiota, N.; Yamada, T.; Inui, H.; Ohkawa, Y.; *Plant Biotechnology* **1998**, *15*, 173-176.

5. *Regulation of Enzymatic Systems Detoxifying Xenobiotics in Plant;* Jepson, J., Holt, D, C., Roussel, V., Wright, S. Y. and Greenland, A. J; Hatzios, K. K., Ed.; Klower Academic Publishers, the Netherlands, **1977**, pp 313-323.

6. *Pesticide Chemistry and Bioscience;* Kreuz, K. and Martinoia, E.; Brooks, G. T. and Roberts, T. R. Eds.; The Royal Society of Chemistry, Cambridge, **1999**; pp 279-287.

7. *Pesticide Chemistry and Biocience;* Ohkawa, H., Imaishi, H., Shiota, N., Yamada, T. and Inui, H.; Brooks, G. T., and Roberts, T. R., Eds; The Royal Society of Chemistry, Cambridge, **1999**, pp 259-264.

8. Winkler, R. G.; Frank, M. R.; Galbraith, D. W.; Feyreisen, R.; Feldmann, K. A.; *Plant Physiol.,* **1998**, *118*, 743.

9. Imajuku, Y., Tsujii, H., Thi, Q. D.and Ohkawa, H.; Ohkawa, H. and Ohkawa Y. Eds; 2nd BRAIN Seminar on Cytochrome P450 and Plant Genetic Engineering, Kobe, Japan, **1999**, pp25-28.

10. Funae, Y.; Obatan N.; Kirigami, *S; Kan Tan Sui*, **1998**, *37*, 91.

11. Inui, H.; Kodama, T.; Ohkawa, Y; Ohkawa, H; *Pesticide Biochemistry and Physiology*, **2000**, *66*.

12. Inui, H., Shiota, N., Motoi, Y., Ido, Y., Ueyama, Y., Inoue, T., Kodama, S., kodama, T., Ohkawa, Y. and Ohkawa H.; Ohkawa, H. and Ohkawa Y. Eds; 2nd BRAIN Seminar on Cytochrome P450 and Plants Genetic Engineering, Kobe, Japan, **1999**, pp44-48.

13. Inui, H.; Ueyama, Y.; Shiota, N.; Ohkawa, Y.; Ohkawa, H.; *Pesticide Biochemistry and Physiology*, **1999**, *64*, 33-46.

14. *Environmental Endocrine Disruptors;* Keith, L. H.; **1997**, John Willey & Sons, Inc., New York, p802.

Chapter 11

Utilizing the Systemic Acquired Resistance Signal Transduction Pathway to Enhance Plant Health

Kay A. Lawton[1], Leslie Friedrich[1], Rebecca Cade[1], Michael Willits[1], Laura Weislo[1], Robert A. Dietrich[1], Michael Oostendorp[2], and John Salmeron[1]

[1]Novartis Crop Protection Agricultural Biotechnology Institute, Research Triangle Park, NC 27709
[2]Novartis Crop Protection, Basle, Switzerland

At Novartis we have identified chemicals that can activate the endogenous disease resistance mechanism known as SAR. One of these, BTH, has been developed as a commercial product that will provide a novel tool to add to the integrated pest management (IPM) systems used in modern agriculture. Preventive application of SAR activators will enhance disease management efforts. We have also identified a gene that is required for response to both biological and chemical activators of SAR. High level expression of the *NIM1* gene in Arabidopsis results in plants with increased health that are responsive to subclinical applications of SAR activators as well as to traditional fungicides.

Plants have evolved the means to respond to attack by potentially pathogenic microbes by constitutively producing pre-formed barriers (e.g. waxy cuticles, thickened cell walls) to microbe ingress as well as by activating inducible disease resistance mechanisms (for reviews see: *1,2*). Disease resistance activation may occur both proximal and distal to the site of attempted penetration by the disease causing organism. We have focused on understanding the signal transduction cascade that results in the activation of the inducible disease defense mechanism known as systemic acquired resistance (SAR). In so doing physical (i.e. morphological), molecular and biochemical changes that occur in the plant at the time of initiation, establishment and maintenance of SAR have been described. We have identified and characterized chemical activators of SAR, one of which is proving to be a useful tool for managing disease in agronomically important crops. We have taken a molecular genetic approach to dissect the SAR signal transduction pathway in *Arabidopsis thaliana*. In so doing we may uncover genes that can be transferred to crop plants to provide broad-spectrum disease resistance. We have also used Arabidopsis to investigate the interactions between SAR and the efficacy of standard fungicides.

Biological Induction of SAR

In nature, necrotrophic pathogens can induce disease resistance against subsequent infection by the same pathogen as well as certain other, unrelated pathogens. This broad-spectrum disease resistance is effective proximal to the initial infection (i.e. local acquired resistance; LAR) site as well as distally (i.e. systemic acquired resistance; SAR). For example, in the well-characterized tobacco mosaic virus (TMV)/tobacco pathosystem, inoculation of a local lesion tobacco host plant with TMV confers systemic resistance in inoculated leaves (LAR) and uninoculated leaves (SAR) against not only TMV but also to certain bacterial and fungal pathogens as well. Pathogen-induced necrosis appears to be a key determinant in the initiation of acquired resistance (AR) in the biological model.

The accumulation of salicylic acid (SA) is critical for the initiation and establishment of SAR. SA levels increase in plants that have been treated with an inducing pathogen. The timing and location of the SA increase is consistent with a role for SA in SAR signaling. In addition, exogenous application of SA results in disease resistance to the same spectrum of pathogens as the biological inducer. The role of SA is further supported by the finding that plants that cannot accumulate SA are unable to mount the SAR response (3).

The accumulation of certain pathogenesis-related (PR) proteins has been shown to be correlated with SAR (4). The temporal and spatial accumulation of the acidic isoforms of the PR proteins are consistent with a role in SAR. Furthermore, transgenic plants that constitutively express high levels of these proteins exhibit enhanced disease resistance against certain pathogens supporting a role for these proteins in disease resistance (5). One of these PR proteins, PR-1, has been shown to provide a tightly correlated molecular marker for the SAR response.

Chemical Activation of SAR

The SAR response can be activated chemically. For a chemical to be considered an SAR activator certain criteria must be met (6). First, there must be no direct antimicrobial activity of the compound *in vitro*. Second, there is a time lag between application of the chemical and establishment of the disease resistance. Third, the resistance induced must be effective against the same spectrum of pathogens as seen with the biological inducer. Finally, the chemical should activate expression of the same *PR* genes that are associated with SAR. As noted above, exogenous applications of the endogenous signaling molecule, SA, results in broad-spectrum disease resistance following a time lag after application. SA also results in the induction of *PR* gene expression and protein accumulation. Synthetic chemicals can also induce SAR. Examples of synthetic chemicals that satisfy the criteria of SAR activators are 2,6-dichloroisonicotinic acid (INA) and benzothiadiazole (BTH).

BTH Activation of SAR Signal Transduction

BTH activation of disease resistance has been characterized in wheat (7), tobacco (8) and Arabidopsis (9). In each of these systems BTH-activated disease resistance satisfies the criteria established to define SAR activators. Further, it has been demonstrated that the accumulation of SA is not required for BTH activated SAR since transgenic tobacco and Arabidopsis that cannot accumulate SA can be activated by BTH.

Additional experiments were carried out in Arabidopsis in order to understand more about the activation of SAR by BTH. Arabidopsis is an appropriate experimental system since biological induction of SAR exhibits the characteristics of increased accumulation of SA and *PR* gene expression that results in broad-spectrum disease resistance (10). The accumulation of the SAR marker gene, *PR1*, is induced both locally (i.e. in infected tissue) and systemically (i.e. in uninfected tissues of an infected plant) by inoculation with virulent or avirulent bacterial pathogens in Arabidopsis. Furthermore, it is possible to conduct experiments investigating the interactions among several stress signaling pathways using Arabidopsis since a number of mutants that are compromised in the ability to transduce various stress signals have been identified. In addition, a number of mutants in genetic resistance (i.e. R-gene mediated) and other disease resistance responses have been identified (for recent reviews see: *2,11*). However, it is clear that BTH activates the SAR signal transduction pathway. An Arabidopsis mutant was identified that cannot activate SAR in response to treatment with inducing pathogens, or the chemical activators, SA and INA (*12*). This mutant, called *nim1* for non-inducible immunity, can produce and accumulate SA, so it is not comprised in SA biosynthesis. The *nim1* mutant is also nonresponsive to BTH (*9*).

Effects of NIM1 on Disease Resistance

A functional NIM1 (non-inducible immunity) protein is required for biological and chemical activation of systemic acquired resistance (SAR) in Arabidopsis. We have isolated the Arabidopsis *NIM1* gene by a map-based cloning strategy (*13*). The protein shares homology with a protein involved in mediating the immune response in mammals and fruit flies. We have engineered transgenic Arabidopsis plants with increased *NIM1* gene expression. These plants exhibit enhanced resistance to pathogen infection. Furthermore, as shown in Table 1, *NIM1* overexpressing plants respond to treatment with subclinical concentrations of BTH and various fungicides, enhancing the capacity of these plants to resist attack by a variety of pathogens. This result is consistent with previous work that demonstrates a contribution of the SAR signal transduction pathway to fungicide efficacy (*14*).

Table 1. Enhanced BTH and fungicide efficacy in NIM1 overexpressing lines

	Fungal growth inhibition (%)		
Treatment	*Wildtype*	*Line 6E*	*Line 7C*
Wettable powder	0	10	14
BTH 0.03 ug/L	100	100	100
BTH 0.003 ug/L	0	100	100
fosetyl 5.0 g/L	7	93	80
$Cu(OH)_2$ 2.0 g/L	0	66	77
metalaxyl 0.0125 g/L	59	76	62

To understand the mechanism by which this resistance is conferred, changes in accumulation of *NIM1* mRNA, other SAR markers (e.g.*PR1*) and other genes involved in disease resistance were assayed. In many, but not all cases, the transgenic lines had elevated levels of SAR-related gene expression. From our results it appears that the NIM1 overexpressing plants are primed to respond faster and with more intensity to pathogen attack.

Since salicylic acid (SA) is critical for the establishment of SAR in response to biological inducers, SA levels in the transgenic plants were measured and crosses were made to NahG plants (which cannot accumulate SA). While SA levels are not elevated in the *NIM1* overexpressers, disease resistance is suppressed in double homozygous plants produced from the NahG crosses. These results indicate that SA accumulation is required for enhanced disease resistance observed in these transgenic plants.

Summary

We have identified chemical activators of the endogenous disease resistance mechanism known as SAR. One of these, BTH, has been developed under the trade name of Actigard™ in the United States. It will provide a novel tool to add to the integrated pest management (IPM) systems used in modern agriculture. Preventive application of SAR activators as solo products and/or used in combination with traditional fungicides will enhance disease management efforts. We have also shown that a functional copy of the NIM1 gene is required for response to both biological and chemical activators of SAR. High level expression of the *NIM1* gene in transgenic plants can contribute to enhanced plant health. Arabidopsis plants with elevated NIM1 show increased health and are responsive to subclinical applications of SAR activators as well as to traditional fungicides.

References

1. Ryals, J. A., U. H. Neuenschwander, et al. Systemic acquired resistance. *Plant Cell* **1996,** *8*, 1809-1819.
2. Maleck, K. and K. Lawton. Plant strategies for resistance to pathogens. *Current Opinion in Biotechnology* **1998,** *9*, 208-213.
3. Gaffney, T., L. Friedrich, et al. Requirement of salicylic acid for the induction of systemic acquired resistance. *Science* **1993,** *261*, 754-756.
4. Ward, E. R., S. J. Uknes, et al. Coordinate gene activity in response to agents that induce systemic acquired resistance. *Plant Cell* **1991,** *3*, 1085-1094.
5. Alexander, D., R. M. Goodman, et al. Increased tolerance to two Oomycete pathogens in transgenic tobacco expressing pathogenesis-related protein 1a. *Proc. Natl. Acad. Sci, USA* **1993,** *90*, 7327-7331.
6. Kessmann, H., T. Staub, et al. Induction of systemic acquired resistance in plants by chemicals. *Annu. Rev. Phytopathol* **1994,** *32* 439-59.
7. Gorlach, J., S. Volrath, et al. Benzothiadiazole, a novel class of inducers of systemic acquired resistance in wheat. *Plant Cell* **1996,** *8*, 629-643.
8. Friedrich, L., K. Lawton, et al. A benzothiadiazole derivate induces systemic acquired resistance in tobacco. *Plant J.* **1996,** *10*, 61 - 70.

9. Lawton, K., L. Friedrich, et al. Benzothiadiazole induces disease resistance in Arabidopsis by activation of the systemic acquired resistance signal transduction pathway. *Plant J.* **1996,** *10,* 71 - 82.
10. Uknes, S., A. Winter, et al. Biological induction of systemic acquired resistance in Arabidopsis. *Mol. Plant Microbe Interact.* **1993,** *6,* 680 - 685.
11. Ji, C., J. Smith-Becker, et al. Genetics of plant-pathogen interactions." *Current Opinion in Biotechnology* **1998,** *9,* 202-207.
12. Delaney, T., L. Friedrich, et al. Arabidopsis signal transduction mutant defective in chemically and biologically induced disease resistance. *Proc Natl. Acad. Sci., USA.* **1995,** *92,* 6602-6606.
13. Ryals, J. K. Weymann, et al. The Arabidopsis *NIM1* protein shows homology to the mammalian transcription factor inhibitor IκB. *Plant Cell* **1997,** *9,* 425-439.
14. Molina, A., M. Hunt, J. A. Ryals. Impaired fungicide activity in plants blocked in disease resistance signal transduction. *Plant Cell* **1998,** *10,* 1903-1914.

Chapter 12

Engineering Herbicide Tolerance with Glutathione Transferases

R. Edwards[1], C. J. Andrews[2], and Ian Jepson[2]

[1]Department of Biological Sciences, University of Durham, DH1 3LE, United Kingdom
[2]Zeneca Agrochemicals Ltd., Bracknell RG42 6ET, United Kingdom

Glutathione transferases (GSTs) are a diverse group of enzymes responsible for detoxifying electrophilic xenobiotics by catalysing their conjugation with the tripeptide glutathione. In plants, GSTs can be divided into four major types based on their sequences with the type I, or phi class, and type III, or tau class being the most abundant and responsible for detoxifying pesticides. In the case of herbicides, GSTs are a major determinant of selectivity between crops and weeds and using a combination of biochemical and molecular approaches large numbers of phi and tau GSTs have now been cloned and characterised. In addition to determining the detoxifying activities of individual GSTs toward different classes of herbicides *in vitro*, it has also been possible to test the functioning of these enzymes in pesticide detoxification *in vivo* by expressing GSTs in transgenic plants.

Glutathione transferases, also known as glutathione *S*-transferases (GSTs, EC 2.5.1.18) catalyse the conjugation of electophillic xenobiotics with the tripeptide glutathione (γ-glutamyl-cysteinyl-glycine). The resulting *S*-linked glutathione conjugates are water soluble and generally non-toxic and in animals

GSTs have a well established role in drug metabolism and detoxification (*1*). Plants also contain GSTs and in common with the soluble GSTs found in animals, the enzymes in plants are composed of two subunits, each typically with a molecular mass in the range 20- 25 kDa (*2*). Although all active soluble GSTs isolated from animals and plants to date are dimers, each subunit has its own active site, which consists of binding domains for glutathione (G-site) and the hydrophobic co-substrate (H-site). The G-site is highly conserved and is involved in the activation of the sulphydryl group of glutathione to the reactive thiolate anion. In contrast, the H-site is very variable, accounting for the diverse range of drugs and pesticides which can serve as GST substrates (*3*). This diversity in catalytic ability is further extended by the diversity in GSTs themselves. In both animals and plants GSTs are encoded by large gene families, giving rise to many different types of subunits. In addition to this genetic diversity, the numbers of GST isoenzymes formed can be further extended due to the ability of GST subunits to dimerise with either identical subunits to form homodimers, or with differing subunits to form heterodimers (*3*).

Plant GSTs and Herbicide Metabolism

In plants GSTs can be divided into four major groupings based on sequence similarities and the structure of the respective genes (*3,4*). Two of these GST types, the zeta and theta, formerly referred to as type II and type IV GSTs respectively (*4*), are also found in animals and show negligible detoxifying activities toward xenobiotics (*3*). The other two types, the phi, or type I GSTs and the tau, or type III GSTs, are specific to plants and are the most commonly encountered GSTs in crops and weeds (*2,3,4*). In some crops, such as maize, phi GSTs predominate, while in soybean and wheat the tau type are most abundant (*3*). Weeds also contain a mixture of phi and tau class GSTs, though in contrast to domesticated crops were GSTs can collectively account for 2% of the total protein present in the leaves, in weed the levels of GST expression are at least 10 times lower (*5*). Significantly, it is this difference in GST abundance, rather than differences in the types of GSTs present which account for differences in the rates of GST-mediated detoxification of herbicides in crops and weeds, which in turn is a primary determinant of herbicide selectivity (*6*).

Thus, with the plant GSTs, we have an extensive family of enzymes which collectively play key roles in determining rates and routes of herbicide metabolism and in dictating herbicide tolerance. As GSTs are also relatively straight forward to purify from plants, it would initially seem straight forward to unequivocally assign specific roles in herbicide detoxification to individual GSTs. However, for the following reasons this has been difficult to achieve.

1. Although GSTs catalyse the *S*-glutathionylation of herbicides there are a number of compounds which will readily undergo such conjugation in the absence of any enzyme. As glutathione is present in plant cells in adequate quantities to support non-enzymic conjugation, there has been some debate on the relative importance of GST-catalysed detoxification reactions notably with chloroacetanilide herbicides in maize (*7*) and fenoxaprop in wheat and grass weeds (*8*).

2. The presence of multiple related GSTs in plants poses problems in purifying individual isoenzymes to complete homogeneity for detailed study. This technical difficulty is compounded by the overlapping specificity of GSTs toward individual herbicides (*9,10*).

3. Environmental stress, plant development and exposure to chemicals, especially herbicide safeners, can all radically alter the profile of GST expression and hence the relative contributions made in herbicide detoxification by individual isoenzymes (*2,11,12*).

All these factors compound the usual difficulties encountered in predicting how enzymes will function in plants *in vivo* based on a knowledge of their kinetic behaviour and regulation obtained from studies carried out *in vitro*. In turn, an understanding of the roles of specific GSTs in herbicide metabolism and selectivity in crops and weeds is important if we are to use the detailed knowledge of structure activity relationships and X-ray crystallographic structural data derived from these enzymes in the rational design of new selective herbicides.

There is therefore a need to develop new approaches to studying the functioning of individual detoxifying enzymes *in vitro* as well as modelling the functioning of these enzymes *in planta*.

Using examples derived from the authors laboratories two case histories where the functioning of GSTs in herbicide metabolism in crop plants have been investigated in detail will be presented. In the first example, individual GSTs present in soybean have been cloned and the recombinant enzymes characterised. This molecular approach was found to be useful due to the complex mixture of closely related GSTs present in whole soybean plants. In the second example, the functioning of individual maize GSTs in detoxifying herbicides has been examined *in planta* using a transgenic approach. Both these lines of research suggest future directions for the characterisation of pesticide detoxification systems in plants. The structures of the herbicides referred to are presented in Figure 1.

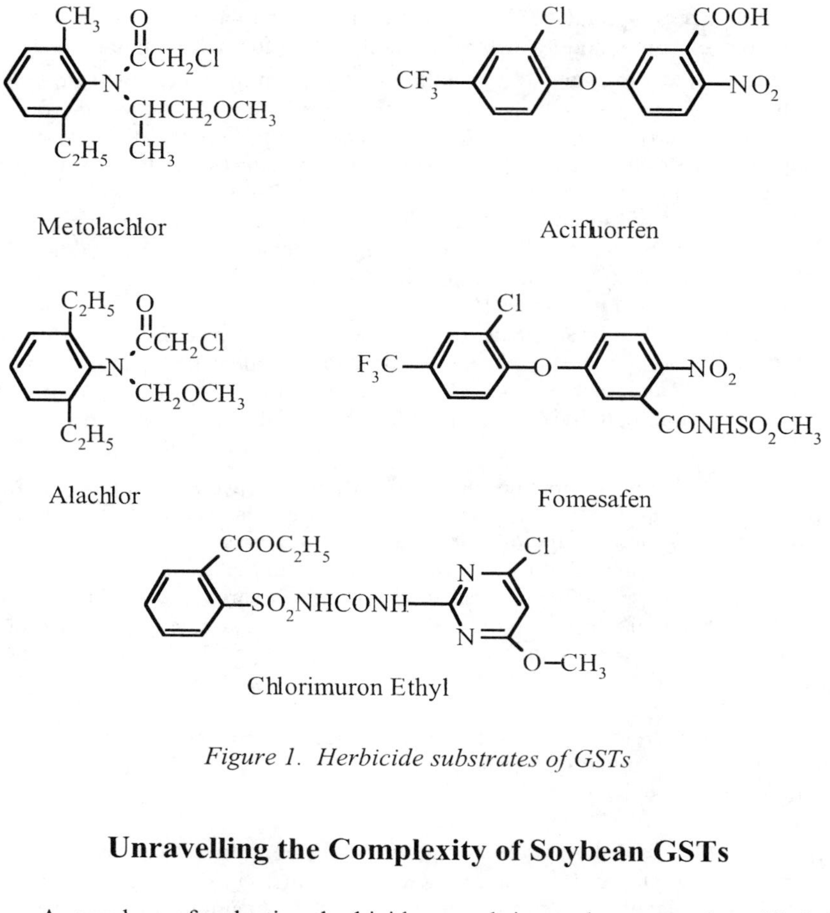

Figure 1. Herbicide substrates of GSTs

Unravelling the Complexity of Soybean GSTs

A number of selective herbicides used in soybean are detoxified by conjugation with homoglutathione (γ-glutamyl-cysteinyl-β-alanine), which replaces glutathione as the dominant thiol in this legume species (*13*). Thus the sulphonylurea chlorimuron-ethyl, the chloroacetanilide metolachlor and the diphenyl-ether herbicides acifluorfen and fomesafen are all rapidly metabolised by homoglutathione conjugation in soybean, this detoxification being a major contributor to the tolerance of the crop to these herbicides (*14*). Surprisingly, in view of the importance of homoglutathione conjugation in herbicide metabolism, little was known about soybean GSTs until recently. A key role for GSTs in the detoxification of chlorimuron-ethyl and the diphenyl ethers was quickly established, as these compounds undergo negligible rates of homoglutathione conjugation in the absence of any enzyme (*15*). Therefore the rapid metabolism observed *in planta* could only be explained as being mediated by GSTs. GST activities toward herbicides were determined in soybean and competing weeds, with the highest activities determined in soybean, consistent with the proposal that GSTs were a primary determinant of herbicide selectivity (*15*).

When the GSTs present in soybean plants and cell cultures were purified by affinity chromatography, at least nine GST subunits could be resolved using reversed-phase HPLC (*16*). A typical separation of the individual GST polypeptides is shown in Figure 2. These GST subunits have very similar molecular masses, with the respective isoenzymes proving very difficult to resolve. This difficulty in isolating individual GSTs led to the use of an alternative approach, in which available DNA and protein sequence data derived from soybean and other GSTs was used to clone several tau class GSTs. These clones were then expressed as fully functional GSTs in recombinant *E. coli* (*13,17*). It was then straightforward to purify large amounts of pure GSTs from the recombinant bacteria. The recombinant GST subunits were first matched to the respective polypeptides isolated from soybean plants. This was achieved by matching the physical and chromatographic behaviour of the recombinant GST with the GST subunits resolved by HPLC (Figure 2). From this analysis it was then possible to identify the respective subunits *in planta* from the HPLC profile (Figure 2). The advantage of using the cloning approach was that it was possible to produce large amounts of GSTs which were not contaminated with the closely related GSTs present in the preparations obtained from plants.

The GST *Gm*GST1, previously identified as a stress inducible protein in soybean, was a tau class GST. Tau class GSTs which showed an unusual substrate dependence in preferred thiol usage when detoxifying herbicides (*13*). With most xenobiotic substrates *Gm*GST1 either showed no preference in its utilisation of glutathione or homoglutathione, or a slight preference for glutathione. However, the diphenyl ether herbicides acifluorfen and fomesafen were only detoxified at appreciable rates in the presence of homoglutathione. Subsequently, a similar thiol dependence was reported for another tau class enzyme cloned from soybean, *Gm*GST2 (*17*), with a related GST showing a preference for homoglutathione when detoxifying chlorimuron-ethyl.(*18*). This kind of detailed information would have been difficult to obtain using the partially purified GST preparations obtainable for soybean plants and the use of recombinant GSTs will be increasingly important if we are to use a knowledge of herbicide detoxifying enzymes to help direct the design of new selective herbicides.

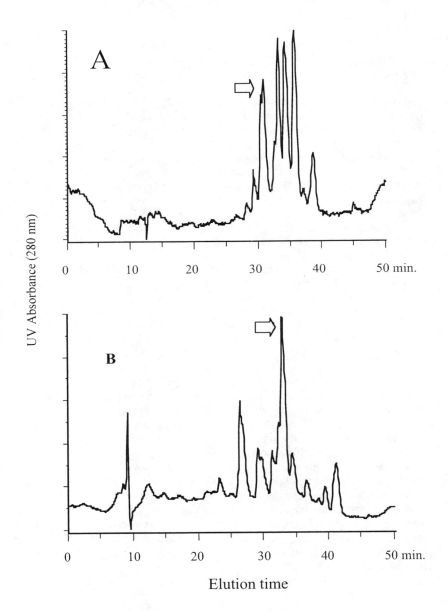

Figure 2. Resolution of GST subunits in (A) soybean cell cultures and (B) soybean seedlings by reversed-phase HPLC. The major GST subunit in seedlings is arrowed in both chromatograms.

Critically Evaluating the Role of Maize GSTs in Herbicide Detoxification

The GSTs of maize are better studied than those of any other plant, yet until recently the relative importance of the individual isoenzymes in herbicide detoxification was unknown. In maize foliage phi GSTs predominate, with ZmGST I, also termed GST 29 being the major subunit present (9). A second phi type GST, termed ZmGST II or GST27, accumulates following treatment with herbicide safeners (9,19). These subunits can dimerise together to form the ZmGSTI-I and ZmGSTII-II homodimers as well as the ZmGST I-II heterodimer. In terms of abundance, these three phi GST isoenzymes are the dominant GSTs determined in safener-treated foliage, though additional isoenzymes composed of phi GST ZmGST III and three tau class GSTs ZmGST V, ZmGST VI and ZmGST VII are also present in lower amounts (10,11). The role of ZmGST I and ZmGST II in the detoxification of the chloroacetanilide herbicides has been of particular interest.

Some studies have argued that chloroacetanilide herbicides are largely detoxified by glutathione conjugation in maize without requiring the intervention of GSTs (7). However, assuming GSTs are involved, the complex combination of GST isoenzymes present makes it difficult to predict whether one form of GST is predominantly responsible, or if detoxification is achieved by differing types of GSTs functioning together. For example, although ZmGST II is more active in detoxifying metolachlor and alachlor than ZmGST I, it is far less abundant. Studies with mutant inbred maize lines showing enhanced sensitivity to alachlor have suggested that ZmGST II is important in determining tolerance to chloroacetanilide herbicides (20). However, a more definitive approach has been to generate transgenic tobacco lines expressing ZmGST I or ZmGST II and then in a series of crosses, generate plants which either express ZmGST I-I, ZmGSTI-II or ZmGST II-II as the active isoenzymes (21).

The coding sequences of ZmGST I and ZmGST II were cloned into the Ti vector pJR1Ri which uses the CaMV 35S promoter and nos terminator and these constructs introduced into tobacco (cv. Samsun) using Agrobacterium tumefacians mediated transformation (21). After selection the lines were then crossed to generate ZmGST I-I, ZmGSTI-II and ZmGST II-II expressing lines, with expression of the maize GSTs monitored by a combination of chromatographic resolution of individual isoenzymes, enzyme assays and the use of specific anti-GST-sera (21).

Spray trials with metolachlor (1400 g/Ha) showed that tobacco lines expressing ZmGST I-II and ZmGST II-II expressing lines showed no visible phytotoxic injury. In contrast the untransformed control plants were very sensitive to metolachlor and were severely affected. Interestingly, plants expressing *Zm*GST I-I showed similar levels of phytotoxic damage to the untransformed controls, even though the transgenic tobacco was expressing significant amounts of the GST. This result demonstrated that the dominant GST in maize may play only a minor role in determining tolerance to chloroacetanilide herbicides, which would be extremely difficult to predict from *in vitro* studies alone. In future the availability of these transgenic models will be an important resource in identifying and understanding the role of detoxification in herbicide selectivity.

Acknowledgements

Robert Edwards acknowledges the support of Zeneca Agrochemicals Ltd and the Biotechnology and Biological Sciences Research Council of the United Kingdom in supporting the research described and thanks Zeneca and the ACS for their generous assistance while attending the 2nd Pan Pacific Congress.

References

1. Hayes, J. D.; Mclellan, L. I. *Free Radical Res.* **1999**, *31*, 273-300.
2. Marrs, K. A. *Annu. Rev Plant Physiol. Plant Mol. Biol.* **1996**, *47*, 127-158.
3. Edwards, R.; Dixon, D. P.; Walbot, V. *Trends Plant Sci.* **2000** (in press).
4. Dixon, D. P.; Cummins, I.; Cole, D. J.; Edwards, R. *Curr. Opin. Plant Biol.* **1998**, *1*, 258-266.
5. Hatton, P. J.; Cummins, I.; Cole, D. J.; Edwards, R. *Physiol. Plant.* **1999**, *105*, 9-16.
6. Hatton, P. J.; Dixon, D.; Cole, D. J.; Edwards, R. *Pestic. Sci.* **1996**, *46*, 267-275.
7. Jablonkai, I.; Hatzios, K. K. *J. Agric. Food Chem.* **1993**, *41*, 1736-1742.
8. Tal, J. A.; Hall, J. C.; Stephenson, G. R. *Weed Res.* **1995**, *35*, 133-139.
9. Dixon, D.; Cole, D. J.; Edwards, R. *Pestic. Sci.* **1997**, *50*, 72-82.
10. Dixon, D. P.; Cole, D. J.; Edwards, R. *Plant Mol. Biol.* **1999**, *40*, 997-1008.
11. Dixon, D. P.; Cole, D. J.; Edwards, R. *Plant Mol. Biol.* **1998**, *36*, 75-87.
12. Jepson, I.; Lay, V. J.; Holt, D. C.; Bright, S. W. J.; Greenland, A. J. *Plant Mol. Biol.* **1994**, *26*, 1855-1866.
13. Skipsey, M.; Andrews, C. J.; Townson, J. K.; Jepson, I.; Edwards, R. *FEBS Lett.* **1997**, *409*, 370-374.
14. Cole, D. J.; Edwards, R. Secondary metabolism of agrochemicals in plants;

In *Agrochemicals and plant protection*; Roberts, T. R., ed. John Wiley and Sons: Chichester, **2000**; pp 107-154.

15. Andrews, C. J.; Skipsey, M.; Townson, J. K.; Morris, C.; Jepson, I.; Edwards, R. *Pestic. Sci.* **1997**, *51*, 213-222.

16. Skipsey, M.; Andrews, C. J.; Townson, J. K.; Jepson , I.; Edwards, R. *Arch. Biochem. Biophys.* **2000** (in press).

17. Andrews, C. J.; Jepson , I.; Skipsey, M.; Townson, J. K.; Edwards, R. *Plant Phys.* **1997** , *113*, 1005.

18. McGonigle, B.; Lau, S. M. C.; Jennings, L. D.; O'Keefe, D. P. *Pestic. Biochem. Physiol.* **1998**, *62*, 15-25.

19. Holt, D. C.; Lay, V. J.; Clarke, E. D.; Dinsmore, A.; Jepson, I.; Bright, S. W. J.; Greenland, A. J. *Planta* **1995**, *196*, 295-302.

20. Rossini, L.; Jepson, I.; Greenland, A. J.; Sari Gorla, M. *Plant Physiol.* **1996**, *112*, 1595-1600.

21. Jepson, I.; Holt, D. C.,;Roussel, V.; Wright, S. Y.; Greenland, A. J. Transgenic plant analysis as a tool for the study of maize glutathione *S*-transferases; In *Regulation of Enzymatic Systems Detoxifying Xenobiotics in Plants*; Hatzios, K. K., ed. Kluwer Academic Publishers, Netherlands, **1997**; pp 313-323.

Chapter 13

Trends in Biocontrol Research on Soilborne Plant Diseases: In Integrated Management Programs on Clubroot Disease of Crucifers

Tsutomu Arie[1,3], Katsuyoshi Yoneyama[2], and Isamu Yamaguchi[1]

[1]Microbial Toxicology Laboratory, RIKEN, Wako 351–0198, Japan
[2]Faculty of Agriculture, Meiji University, Kawasaki 214–0033, Japan
[3]Present address: Faculty of Agriculture, Tokyo University of Agriculture and Technology, Fuchu 183–8509, Japan

Biocontrol (biological control) research on plant diseases faces a new challenge. Since it might be environmental friendly, biocontrol was expected to be an outstanding method for plant protection. However, after two decades of laborious studies carried out by many researchers, it has been shown that biocontrol alone could not be a complete replacement for management strategies currently employed. The instability of the biocontrol agent is one of the reasons for this. Therefore, researchers are now groping for a new way to exploit the potential of biocontrol maximally by utilizing it in integrated pest management. Furthermore, analysis of biocontrol mechanisms has become the major interest of research, which may contribute to the future innovation of management programs for crop production.

Introduction

The population explosion coming early in the 21st century is predicted and development of a sustainable agriculture system to supply enough safe crops is urgently in needed (1). Biocontrol (biological control) of plant diseases has been expected to meet this objective, because it can be safer for humans, crops, and the environment, and moreover may allow reduction of chemical pesticides used worldwide (2). Cook and Baker (3) introduced to the world the possibility of biocontrol. They showed "What biocontrol" can do appeared in their book in 1974, and in 1983 Baker and Cook (4) again showed their idea by asking "How biocontrol?" About 20 years has passed since the books appeared and during that period, many researches have tried to obtain useful biocontrol agents through screening natural microorganisms. The target diseases were mainly soilborne diseases, and many microorganisms, such as *Pseudomonas* spp. (5-10), *Bacillus* spp. (11), *Erwinia* spp. (12), *Fusarium* spp. (13, 14), *Trichoderma* spp. (15), and *Gliocladium* spp. (15) etc., have been reported as candidates. However, most of them have never been used practically (16). Most of the candidates have been inconsistent and/or unsatisfactory and ecologically not stable enough in the new environment. These findings suggest that researchers should be more careful in introducing biocontrol agents into the environment, especially into the soil (2, 21, 22). The examples of the biocontrol agents registered as bio-fungicides together with several candidates which are in the final stage of testing for registration in Japan are listed in Table I.

It is important to understand the mechanisms of action of biocontrol agents in order to use their maximum potential. The known mechanisms of action can be summarized as follows (2, 4, 23):

• Competition
• Antibiotic production
• Parasitism/ predation
• Induced resistance/ cross protection.

As molecular and biochemical techniques develop, the mechanism of action and the stability of the biocontrol agent which have not been clarified will finally be made clear.

The other approach is to maximize the effect of biocontrol agents by using

Table I. Biocontrol Agents Registered as Bio-Fungicides and Those at the Final Stage of Testing for Registration in Japan

Agent	Target Disease	Causal Pathogen	Year of Regist-ration	Reference
Trichoderma lignorum	Stem rot and damping-off of tobacco	*Sclerotium rolfsii* & *Rhizoctonia solani*	1956 & 1968	
Agrobacterium radiobacter strain 84	Crown gall of rose	*A. tumefaciens*	1989	*17*
Non-Pathogenic *Erwinia carotovota*	Soft rot of Chinese cabbage	*E. carotovota*	1997	*12*
Bacillus subtilis IK-1080	Gray mold of tomato and egg plant	*Botrytis cinerea*	1998	*18*
Attenuated Zucchini Yellow Mosaic Virus	Mosaic of cucumber	Zucchini Yellow Mosaic Virus	FS[a]	
Pseudomonas sp.	Bacterial grain rot of rice	*Burkholderia glumae*	FS	*7*
Ralstonia solanacearum	Bacterial wilt of tomato	*R. solanacearum*	FS	*19*
Taralomyces flavus	Anthracnose and powdery mildew of strawberry	*Colletotricum acutatum* & *Sphaerotheca humuli*	FS	
Trichoderma harzianum	Rhizoctonia patch of Japanese lawngrass	*Rhizoctonia solani*	FS	
Aphanomyces quisqualis	Powdery mildew of cucumber, strawberry, rose	*Sphaerotheca* spp.	FS	

SOURCE: Information from Reference *20*.

[a] FS: Agents at the final stage of test for registration.

them in an integrated management strategy. An integrated management program is composed of several techniques, such as field sanitation, field diagnosis, agrochemical, biocontrol, disease-resistant cultivar etc., and each will supplement the shortcomings of each technique. Integrated management programs seem to have almost no risk such as incapacitation of the program, since it is not dependent on a single tool, and it is also possible that the adverse effect of each method on the environment can be reduced. As commented by Larkin (2): because of its nature biocontrol should be perceived as a complementary tool for an integrated management program.

A great number of books and reviews concerning biocontrol and integrated management programs of plant diseases have been published offering general information (e.g. 2, 3, 4, 21, 22). In this chapter, we present the outline of some new traits in biocontrol-research in Japan by showing the examples on clubroot of cruciferous plants.

Clubroot of Crucifers

Soilborne clubroot disease caused by *Plasmodiophora brassicae* Woron. is a major threat to *Brassica* farming (cabbage, Chinese cabbage, turnip, broccoli, cauliflower, and horseradish) worldwide. The pathogen of this disease has diverse pathogenic varieties and remains in the soil in the form of resting spores for a long period that makes the control of this disease very difficult (24, 25). Conventionally, a fungicide, pentachloronitrobenzene (PCNB; quintozene), in powdery formulations have been applied into the infested soil (a.i. ~ 40 kg/ ha; a.i. ~ 1,250 t in 1996 in Japan) (26). The prolonged use of PCNB has resulted in an accumulation of PCNB and its major metabolite pentachloronitroaniline (PCA) in soil. Both are physicochemically stable and may cause environmental problems in the future. Although several new fungicides, such as flusulfamide and fluazinam, have been introduced to the market, the efficacy of these newly developed fungicides seems lower than expected, especially when field conditions are different (19). This suggested the need to develop alternative control systems.

Many attempts have been made for biocontrol of clubroot. Gramineous plants, leguminous plants, and radish are reported to be used as cleaning crops (*i.e.*, decoy plants) (25, 27). Soil fungi and bacteria, antagonistic to the pathogen, have been screened (*28-31*).   A soil-amendment, composed of lime-grains and a bacterium *Bacillus* sp., was reported to be effective in the control of the disease (*11*).

Root Endophytic Fungus showing Stable Biocontrol Activity of Clubroot

An ascomycete, *Heteroconium chaetospira* (Grove) Ellis M4007, was obtained from the rhizosphere of healthy Chinese cabbage (*B. campestris*) plants and was shown to have a good suppressive effect on clubroot (*31*).   This biocontrol agent was inoculated into the root system of Chinese cabbage by sowing seeds in the M4007-colonized nursery bed composed of peat mosses and mycelia of the fungus.   M4007 colonizes the root surface and the inner cortical tissues of the root of the Chinese cabbage during plant growth.   It could be re-isolated from the root system 3 months after the inoculation, indicating that the strain is endophytic (able to colonize inner plant tissues– mutualistic). Although the biochemical mechanisms of action of this fungus are not known, *H. chaetospira* M4007 is expected to promote induced resistance into Chinese cabbage.

Utilization of endophytic microorganisms seems to be an added method in solving the problem of instability of biocontrol.

Clubroot Suppression by *Phoma glomerata* and its Non-Antifungal Product, Epoxydon

An ascomycete, *Phoma glomerata* (Corda) Wollenw. & Hochapfel JCM9972[1], was obtained as a biocontrol agent against clubroot (*32*).   During the investigation of mechanisms of action of the fungus, *P. glomerata* JCM9972, it was shown to produce an active compound, epoxydon, 5-hydroxy-3-(hydroxymethyl)-7-oxabicyclo[4.1.0]hept-3-en-2-one (*32*). Although epoxydon exhibits almost no anti-microbial activity against many species of pathogenic

---

[1] JCM: Japan Collection of Microorganisms, RIKEN, Wako, Saitama 351-0198, Japan.

fungi and bacteria, such as *Fusarium oxysporum* Schlechtend.: Fr., *Alternaria mali* Roberts, *Botrytis cinerea* Persoon : Fries, *Burkholderia caryophylli* Yabuuchi, *et al.*, and *Xanthomonas campestris* (Pammel) Dows, it could prevent clubroot of crucifers with irrigation of 100 µg/ml solution (30 ml/ 200 g infested soil).

Epoxydon was initially purified from cultured broth of *Phoma* sp. by Closse *et al.* (*33*) as an anti-tumor substance, and later, it was reported to have anti-auxin activity (*34*). Considering the fact that auxin plays an important role in gall formation in clubroot of crucifers (*35, 36*), the clubroot suppression activity by epoxydon might be due to its anti-auxin activity. Actually, several known anti-auxins, *e.g.*, 2,3,5-triiodobenzoic acid, have shown significant effect on clubroot suppression (*32*).

P. glomerata JCM9972 and its product, epoxydon, shows a new mechanism of action (suppression of the symptom-development of clubroot). Since the effect is not due to anti-microbial activity, it could be environmentally compatible and might not produce resistant strains of the pathogen.

Integrated Management Programs for Clubroot

Methods listed in Table II illustrate the integrated management programs for clubroot of crucifers. To construct an integrated management program, coping well with each field, field diagnoses of soil properties and the presence of pathogens in soil and/or nursery seedlings must be an important initial step. Current immunology provides rapid, specific, and sensitive tools for the detection of soilborne pathogens (*46*). In the case of *P. brassicae*, polyclonal antibodies against the resting spores have been prepared, and immunofluorescence assays (IFA) and enzyme-linked immunosorbent assays (ELISA) were recommended to be helpful in detecting resting spores in soil (*40, 41*). Tsushima (*37*) proposed a biointensive-integrated management program for clubroot, emphasizing the importance of field diagnosis and field sanitation as well as biocontrol in the program.

The integrated management program can be improved by reducing the possible hazards from chemical and/or biological pesticides in the environment. The amount of the residual pesticides should be monitored correctly, and then remediation procedures should be applied (Table II). For example, PCNB

Table II. Methods for Integrated Management Program on Clubroot

Method	Reference
Field sanitation	
Cleaning agricultural equipment	*37*
Control soil water	*38*
Nursery paper-pot	*38*
Crop rotation	*38*
Soil property analysis	*37*
Soil amendment	
CaCO$_3$	*38*
Coral Reef Rock	*39*
Zn and B	*25*
Field diagnosis	
Indicate plant	*24, 25*
Immunological method	*40, 41*
Occurrence forecasts	
Remote-sensing	*38*
Chemical fungicide	
PCNB	
Fluazinum	*42*
Flusulfamide	*43*
Biocontrol	
Bacillus sp.	*11*
Heteroconium chaetospira	*31*
Phoma glomerata	*32*
Decoy plants	*25, 27, 37*
Resistant varieties	
CR series	*38*
Monitoring of fungicides in the environment	
GC/MS	*44, 45*
Immunological method	
Biodegradation of PCNB in the environment	
Microbes	*44*
Rhizo-microbes	*45*

accumulated in the field soil is easily detected by GC/MS assay, and microbes such as *Pseudomona*s spp. degrade it (*44, 45*).

Biocontrol Research in the Future

In this chapter, we have summarized the current situation of biocontrol in Japan by showing examples on clubroot disease.   The application of an integrated management program on clubroot should provide more precise control in infested fields.   Optimized integrated management programs including biocontrol means will be widely applied to other soilborne diseases in the near future.   To achieve these objectives, we would like to emphasize the necessity of accumulation of basic knowledge on biocontrol such as mode of action of biocontrol and stability of biocontrol agents.   To date the biocontrol research has been mainly carried out by plant pathologists, but now is the time for other scientists such as pesticide scientists and plant scientists to play their important role in order to accelerate research in the field of biocontrol.

On the other hand, a new trend for application of biocontrol technology onto post-harvest diseases is now in progress.   It is expected that the application of biocontrol to the post-harvest diseases will be popular because the stabilization of the biocontrol agent is easier due to the simple biota and that it may result in agricultural products with reduced pesticide residues.

Continuous efforts on biocontrol research will bring about a stable supply of agriculture crops by applying biocontrol methods more widely to soilborne and post-harvest diseases.

Acknowledgments

The authors would like to express their thanks to Dr. T. Suzui, BRAIN, Tokyo, Japan and Dr. S. Tsushima, Tohoku National Agricultural Experiment Station of MAFF, Fukushima, Japan for offering valuable information.

References

1.   Evans, D. A. *Pesticide Chemistry and Bioscience: The Food-Environment Challenge;* Royal Society of Chemistry: Cambridge, UK, 1999; p 3.
2.   Larkin, R. P.; Roberts, D. P.; Gracia-Garza, J. A. *Fungicidal Activity;* John Wiley & Sons: New York, NY, 1998; p 149-191.
3.   Bɛ   , K. F.; Cook, R. J. *Biological Control of Plant Pathogens;* APS: St. Pauı, MN, 1974.

150

4. Cook, R. J.; Baker, K. F. *The Nature and Practice of Biological Control of Plant Pathogens;* APS: St. Paul, MN, 1983.

5. Arie, T.; Namba, S.; Yamashita, S.; Doi, Y.; Kijima, T. *Ann. Phytopathol. Soc. Jpn.*, **1987**, *53*, 531-539.

6. Homma, Y.; Suzui, T. *Ann. Phytopathol. Soc. Jpn.*, **1989**, *55*, 643-652.

7. Tsushima, S.; Torigoe, H. *Plant Protection*, **1991**, *45*, 91-95.

8. Tsuchiya, K. *PSJ Biocont. Rept.*, **1994**, *4*, 35-44.

9. Furuya, N.; Yamasaki, S.; Nishioka, M.; Shiraishi, I.; Iiyama, K.; Matsuyama, N. *Ann. Phytopathol. Soc. Jpn.*, **1997**, *63*, 417-425.

10. Murakami, K.; Kanzaki, K.; Okada, K.; Matsumoto, S., Oyaizu, H. *Ann. Phytopathol. Soc. Jpn.*, **1997**, *63*, 432-436.

11. Kijima, T., Namai, K., Goma, H. *Soil Microorganisms*, **1998**, *52*, 65-71.

12. Kikumoto, T. *PSJ Soilborne Dis. Rept.*, **1998**, *19*, 87-98.

13. Ogawa, K.; Komada, H. *Ecology and Management of Soilborne Plant Pathogens;* Parker, C. A. *et al.*, Eds; APS: St. Paul, MN, 1985.

14. Katsube, K.; Akasaka, Y. *Ann. Phytopathol. Soc. Jpn.*, **1997**, *63*, 389-394.

15. Suzuki, G. *PSJ Biocontrol Rept.*, **1999**, *6*,1-9.

16. Komada, H. *Reports on Pesticide-Biotechnology Techniques*, March, 1998; p 1-12.

17. Tomono, K. *PSJ Biocont. Rept.* **1994**, *4*, 16-23.

18. Chida, S. *PSJ Biocont. Rept.* **1999**, *6*, 85-91.

19. Aino, M.; Maekawa, Y.; Mayama, S.; Kato, H. *Diversity and Use of Agricultural Microorganisms;* NIAR, MAFF: Tsukuba, Japan, 1997; p 66-76.

20. Suzui, K. *PSJ Biocontrol Workshop Report*, **1999**, *6*, 98-106.

21. Jacobson, B. J. *Annu. Rev. Phytopathol.* **1997**, *35*, 373-391.

22. Hoitink, H. A. J.; Boehm, M. J. *Annu. Rev. Phytopathol.* **1999**, *37*, 427-446.

23. Hyakumachi, M. *J. Pesticide Sci.*, **1998**, *23*, 422-426.

24. Tanaka, S.; Yoshihara, S.; Ito, S.; Kameya-Iwaki, M. *Ann. Phytopathol. Soc. Jpn.*, **1988**, *63*, 183-187.

25. Dixon, G. R. *Tohoku National Agricultural Experiment Station* **1999**, *15*, 8-10.

26. *Annual Handbook on Pesticides 1997;* JPPA: Tokyo, Japan, 1997.

27. Murakami, H.; Tsushima, S.; Akimoto, T.; Shishido, Y., *Phytopathology*, **1999**, *89*, S54-55.

28. Djatnika, I. *Cruciferae Newsletter*, **1991**, *14-15*, 142.

29. Einhorn, G. H.; Bochow, H.; Huber, J.; Krebs, B. *Archiv. für Phytopathologie und Pflanzenschutz*, **1991**, *27*, 205-208.

30. Elsherif, M.; Grossmann, F. *Journal of Plant Disease and Protection*, **1991**, *98*, 236-249.

31. Narisawa, K; Tokumatsu, S.; Hashiba, T. *Plant Pathol.* **1998**, *47*, 206-210.
32. Arie, T.; Kobayashi, Y.; Okada, G.; Kono, Y.; Yamaguchi, I. *Plant Pathol.* **1998**, *47,* 743-748.
33. Closse, A.; Mauli, R.; Sigg, H. P. *Helvetica Chimica Acta,* **1966**, *25*, 204-213.
34. Sakai, R.; Sato, R.; Niki, H.; Sakamura, S. *Plant and Cell Physiology*, **1970**, *11*, 907-920.
35. Raa, J. *Physiologia Plantarum*, **1971**, *25*, 130-134.
36. Ludwig-Mueller, J.; Epstein, E.; Hilgenberg, W. *Physiologia Plantarum*, **1996**, *97*, 627-634.
37. Tsushima, S. *Tohoku National Agricultural Experiment Station*, **1999**, *15*, 27-29.
38. Komada, H. *Soilborne Diseases of Vegetables;* Takii Seeds: Kyoto, Japan, 1998; p 95-100.
39. Arie, T.; Namba, S.; Yamashita, S.; Doi, Y. *Ann. Phytopathol. Soc. Jpn.* **1985**, *51*, 102.
40. Arie, T.; Namba, S.; Yamashita, S.; Doi, Y. *Ann. Phytopathol. Soc. Jpn.*, **1988**, *54*, 242-245.
41. Wakeham, A. J.; White, J. G. *PMPP*, **1996**, *48*, 289-303.
42. Matsuo, N.; Suzuki, K. *J. Pesticide Sci.*, **1995**, *20*, 129-136.
43. Shimotori, H.; Yanagida, H.; Enomoto, Y.; Igarashi, K.; Yoshinari, M.; Umemoto, M. *J. Pesticide Sci.*, **1996**, *21*, 31-35.
44. Tamura, K.; Hasegawa, Y.; Kudo, T.; Yamaguchi, I. *J. Pesticide Sci.*, **1995**, *20*, 145-151.
45. Yamaguchi, I.; Nakashita, H.; Arie, T.; Kobayashi, Y.; Someya, K.; Tamura, K.; Kono, Y.; Okada, G. *Asian Network on Microbial Researches;* Gagjah Mada Univ.: Yogyakarta, Indonesia; p 359-370.
46. Arie, T. *J. Pesticide Sci.*, **1998**, *23*, 349-356.

Chapter 14

Status and Safety of Biotech Crops

James D. Astwood and Roy L. Fuchs

Monsanto Company, 700 Chesterfield Village Parkway North, St. Louis, MO 63198

Almost 100 million acres of biotech crops, especially corn, soybean, cotton, canola and potatoes, were grown worldwide in 1999. Most were modified to confer agronomic traits such as insect protection, herbicide tolerance or virus resistance. Many benefits are being realized: including less farmer reliance on traditional pesticides, improvements in practices to achieve environmentally sustainable farms, and enhanced food and environmental safety. New nutritional benefits, such as oils enriched in β-carotene or oleic acid, as well as biotech crops which produce valuable pharmaceuticals, are beginning to appear. Food and environmental safety has been assessed according to international risk assessment guidelines, albeit there are distinct procedures in various jurisdictions globally. Risk assessment focuses on the properties of the introduced traits, the genes which confer the traits, and the resulting crop. To date, the genes conferring these traits have been established as safe on the basis of their history (i.e. prior existence in foods and the environment), the well understood function and similarity of the encoded proteins to dietary proteins, and the lack of animal or environmental toxicity of the encoded proteins. Food derived from biotech crops have been assessed to be as safe and nutritious as traditional foods and crops.

Introduction

Significant expansion in the utilization of genetically modified crops has been realized since their introduction in 1996. By the crop year 1999, almost 100 million acres of biotech crops were grown, where soybean, corn, cotton, canola and potato represented 99% of the plantings, although 14 other crops were grown to a lesser but increasing extent (*1*). In 1999, biotech crops in the U.S. represented an expanding proportion of the total crop acreage: approximately 50% of soybeans, 33% of corn, 55% of cotton and 4% of potatoes had biotech traits - mainly insect protection or herbicide tolerance. Likewise, Canada (55% of canola and 20% of soybeans), China (10% of cotton) and Argentina (greater than 75% of soybeans) were also significant producers of biotech crops. The rapid adoption of biotech crops has been driven by economic and environmental benefits that are highly recognized by farmers, yet remain obscure to the general public where knowledge and information has been low and concerns about safety and consequences may be increasing. More recently, tangible environmental and potential public health benefits are being identified and communicated to consumers. Likewise, biotech crops with nutritional enhancements and direct benefits to consumers have begun to enter the marketplace or are under development.

Biotech Crops and Pesticide Use

Traits in the first biotech crops have focused on pest protection and herbicide tolerance. A biotech approach should in principle lead to the reduction in farm use of chemical crop protection products with increasing adoption of biotech crops conferring pest protection, such as those utilizing BT crystal proteins derived from *Bacillus thuringiensis*. Corn, cotton, and potato crops expressing BT genes to control lepidopteran and coleopteran insects were developed because of the historical safety of microbial BT sprays and the well described effectiveness of BT proteins. A dramatic example of the reduction of chemical crop protection products has been described for BT cotton which was designed to combat pink bollworm, cotton bollworm and tobacco budworm. Grower applications of insecticides to BT cotton fields averages two or less per year, whereas traditional cotton requires five to eight applications. It was recently estimated that over two million pounds of insecticides have been eliminated from use in cotton fields as a result of BT cotton (*2*). Table I illustrates the types and quantities of pesticides saved.

Table I. Decline in Insecticide Use in US Cotton.

Insecticide (Trade name)	Reduction (thousands of pounds)
Amitraz (Ovasyn)	42
Cyfluthrin (Baythroid)	35
Cypermethrin (Ammo)	81
Deltamethrin (Decis)	-11
Esfenvalerate (Asana)	19
Lambdacyhalothrin (Karate)	58
Methomyl (Lannate)	156
Profenofos (Curacron)	1,014
Spinosad (Tracer)	-19
Thiodicarb (Larvin)	665
Tralomethrin (Scout)	4
Zeta-cypermethrin (Fury)	-1
TOTAL (Net)	2,043

Note: (-) indicates an increase rather than reduction.

Source: National Center for Food and Agricultural Policy, 1999.

In addition to cotton, pesticide use may also be reduced for insect protected corn varieties. Adoption of herbicide tolerant crops, such as glufosinate tolerant and glyphosate tolerant canola, corn and soybeans, may also lead to greater use of herbicides with more environmentally neutral profiles than some current herbicides. Preliminary studies of traditional corn varieties where traditional weed control practices were used versus herbicide tolerant corn varieties suggest that less herbicide contamination of ground water tends to occur in watersheds where herbicide tolerant corn varieties are widely adopted (Gustafson, D., Monsanto Co. unpublished data).

Fumonisins

Insect tolerant crops may also yield improvements in food and feed safety. An example under investigation is the accumulation of fumonisin, a mycotoxin associated with a variety of health complaints that were first described as "moldy corn poisoning" in 1850 (3). Fumonisin accumulation is due to fungi of the genus *Fusarium* which causes plant disease in corn, especially when insect damage such as that caused by European corn borer (*Ostrinia nubilalis*), creates opportunities for infection at wound sites. In insect protected corn varieties, substantial reductions in both *Fusarium* ear rot and in overall fumonisin levels have been observed - a survey of corn varieties in 1997 showed that mean total fumonisin levels were 16.5 µg/g for nontransgenic corn and 2.1 µg/g for insect protected corn, including those expressing the CryA(b) gene or the Cry9(c) genes from *Bacillus thuringiensis* (4).

Nutritionally Enhanced Biotech Crops

Biotech crops with improved quality, processing characteristics and nutritional enhancements, have begun to reach the marketplace. Indeed, the first biotech crop approved by the U.S. agencies was the delayed ripening tomato (FlavrSavr™) which provides improved storage and handling characteristics due to slower ripening processes. The delayed ripening trait was created by anti-sense DNA technology that "switches off" endogenous polygalacturonase that otherwise would soften the fruit more quickly (5).

Recently, modifications of endogenous oil quality and the introduction of enhanced nutritional quality (such as vitamin enrichment) have begun to be realized. Soybeans are now available on the market in which the level of oleic acid has been changed from approximately 24% of the total oil to more than 80%. High oleic acid soybean oil provides an opportunity to substitute saturated fatty acids produced by hydrogenation of vegetable oils with a stable naturally mono-unsaturated fatty acid that is associated with cardiovascular benefits (6).

Vitamin A deficiency has been identified by the World Health Organization to be a significant concern in developing nations, especially those found in southeast Asia where consumption of food containing animal fats (a usual source of vitamin A) is low (7). As many as 250 million children have been estimated to have sub-clinical signs of vitamin A deficiency, which leads to night blindness and other ailments. A recent solution includes the development of biotech crops such as high β-carotene rice (8) and canola oil (9) which can provide crop-grown alternatives to overcome the shortfall (as much as 70%) in the recommended daily intake of vitamin A by delivering precursor β-carotene.

Production of pharmaceuticals which are rare or difficult to manufacture can now be accomplished in plants. Advantages of biotech crops over other gene expression systems include a reduced risk of human pathogen contamination (such as by viruses), the ability of plant systems to affect appropriate post-translational processing of proteins and low cost. Biologically active human somatotropin, which is used in the treatment of hypopituitary dwarfism in children, has recently be produced in tobacco chloroplasts engineered to express the hSt gene (10).

Food Safety Assessment

In 1996, a joint report from an expert consultation sponsored by the World Health Organization (WHO) and the Food and Agricultural Organization (FAO) of the United Nations concluded that "biotechnology provides new and powerful tools for research and for accelerating the development of new and better foods" (*11*). The FAO/WHO expert consultation also concluded that it is vitally important to develop and apply appropriate strategies and safety assessment criteria for food biotechnology to ensure the long-term safety and wholesomeness of the food supply.

Following these criteria, foods derived from biotechnology have been extensively assessed to assure they are as safe and nutritious as traditional foods. All foods, independent of whether they are derived from biotech crops or traditionally bred plants, must meet the same rigorous food safety standard. Numerous national and international organizations have considered the safety of foods derived from biotech crops. They have concluded that the food safety considerations are basically of the same nature for food derived from biotech crops as for those foods derived using other methods like traditional breeding.

This concept of comparing the safety of the food from a biotech crop to that of a food with an established history of safe use is referred to as "substantial equivalence"(*12, 13*). The process of substantial equivalence involves comparing the characteristics, including the levels of key nutrients and other components, of the food derived from a biotech crop to the food derived from conventional plant breeding. When a food is shown to be substantially equivalent to a food with a history of safe use, "the food is regarded to be as safe as its conventional counterpart" (*11*). A FAO/WHO expert consultation in 1995 concluded that "this approach provides equal or greater assurance of the safety of food derived from genetically modified organisms as compared to foods or food components derived by conventional methods" (*11*).

Protein Safety

Central to the evaluation of the safety of a resulting food, specific protein or derived product, has been a thorough assessment of the genes and, where appropriate, the encoded proteins for all transgenes. One important hazard would be the introduction into a food crop of a protein that could cause an illness or adverse reaction, such as a highly toxic protein or an allergen.

Usually when a gene is chosen for transformation into a crop, the encoded protein has been well characterized in terms of function (mechanism of action, evolutionary heritage, physicochemical properties, etc.). This information has been extensively evaluated during the development of biotech crops such as

NewLeaf™ potato, (*14*) RoundupReady™ soybeans (*15*) and YieldGard™ corn (*16*). An important consideration in protein safety is whether or not the protein can be established to have been used or eaten previously - is there a history of safe use? BT proteins, for example, have been safely used in microbial pesticide formulations for more than 30 years. There existed a substantial toxicological database (including acute, subchronic and chronic animal feeding trials; and an acute human feeding trial) for microbial BT pesticides which corroborated the safe profile of BT proteins for use in biotech crops (*17*).

Protein digestibility studies (*in vitro*) to establish physicochemical similarity to ordinary dietary proteins, bioinformatics screens for known allergens and toxins, and *in vitro* allergenicity studies have been used to confirm protein safety. A full evaluation of the protein's function may yield important insights with respect to safety, either based on scientific first principles, or as determined experimentally. For example, certain enzyme activities could be unsafe if the activity results in the production of toxic chemicals, a property which can be anticipated by careful characterization of the enzymology of the introduced protein. The enzymology of CP4 EPSPS, which confers tolerance to glyphosate, has been well described (*15*) and would not be expected to create toxic secondary metabolites. In addition, endogenous EPSPSs are widely distributed in all food crops and therefore present no new hazard and risk.

Lack of protein toxicity is confirmed by evaluating acute oral toxicity in mice or rats (*18*). This study is typically a two week program in which the pure protein is fed to animals at doses which should be 100 to 1,000 times higher than the highest anticipated exposure via consumption of the whole food product containing that protein. Table II summarizes the data from several acute oral toxicity studies. Although these studies were designed to obtain LD_{50}'s, in fact no lethal dose has been achieved for these proteins (*14, 16, 19-21*).

Table II. Lack of Acute Oral Toxicity of Proteins Introduced into Crops

Protein	Crop	No Observed Effect Level (mg/kg)
Cry1A(b)	corn	4000
Cry1A(c)	cotton, tomato	4200
Cry3A	potato	5200
CP4 EPSPS	soybean, cotton, canola	572
GUS	soybean, sugarbeet	100
NPTII	potato, cotton, tomato	5000

Special Cases

In some cases, the data may suggest that there are unique or unresolved issues associated with a particular protein. Thus, additional studies should be employed. These may include sub-chronic animal feeding studies in avian, ruminant and mono-gastric species. The goal of this type of study (e.g. 90 day rat) would be to confirm that there were no unexpected toxicities associated with the foods derived from biotech crops relative to traditional foods. Doses should represent 10-100 times the anticipated real exposures to these foods. These studies might be anticipated for a crop where a lectin or other suspected toxin had been introduced (22) or where the function of the protein may lead to altered biochemical composition of the resulting whole food.

Case-specific *in vitro* toxicology studies could also be envisaged where the function of a protein may suggest potential pharmacologic activity. For example, a protein that is highly homologous to neurotoxins could be evaluated using *in vitro* and *in vivo* neurological approaches. These studies might be justified only where pharmacology and function indicate a potential concern (23) since there are very few proteins known to exert such effects when presented orally (18).

Allergy

A significant hazard that has been assessed for all biotech crops is the possibility of accidental transfer of existing allergens from one crop to another. The US FDA had proposed a system for evaluating this possibility in 1992 (24). In 1995, the Food and Agricultural Organization Technical Consultation on Food Allergies reported there are few potential allergens in the food supply (11, 12, 25). Over 90% of these allergenic foods result from sources such as fish, peanuts, soybeans, milk, eggs, crustacean, wheat, and tree nuts. Nevertheless, sensitive individuals can experience a reaction from exposure to only trace amounts of allergenic foods. All proteins/genes introduced into genetically enhanced foods have been assessed for allergenic potential early in the development process using internationally accepted experimental and clinical allergy testing methods (11, 25, 26).

Proteins are assessed by evaluating the source of the gene (was the gene obtained from highly allergenic foods, such as Brazil nuts(27), for example) and by evaluating the physical characteristics of the protein to ensure that they show no similarities to known allergenic proteins. Genes obtained from sources known to cause allergy are presumed to encode allergens unless proven otherwise using clinical testing. These procedures are the same tests used by physicians to diagnose and treat food allergy in patients (26). They include laboratory allergy testing, diagnostic skin prick tests, and clinical food challenges. All proteins undergo a further review of detailed items such as whether the new protein has an amino acid sequence similar to known allergens, is resistant to digestion, is heat stable, and is abundant - all factors which contribute to allergenicity of proteins (28). In addition to the protein, the genetically enhanced crop itself should be tested in diagnostic procedures on a case-by-case basis (29).

Substantial Equivalence

The principle strategy for the assessment of substantial equivalence of biotech crops has been to evaluate the safety profile of the introduced traits (usually proteins) and to evaluate the biochemical composition of resulting whole food or feed. Compositional analyses should be performed by sampling food (often grain) grown in a variety of geographies to assess genotype by environment interactions and to understand natural variability in measured parameters. The goal has been to understand whether the composition of the new food or feed falls within the generally accepted definition or specification compared to a traditional variety. For example, the composition of insect protected corn has been compared to traditional corn as illustrated in Table III. The evaluation of composition focuses on macronutrients, vitamins and minerals, but may also include an evaluation of relevant anti-nutrients or toxins such as endogenous phytoestrogens, protease inhibitors and lectins (30).

Agronomic Equivalence

In addition to a demonstration of substantially equivalent composition, further agronomic evaluation of the biotech crop is necessary to establish that there are no unexpected biological effects of the introduced trait. While compositional assessments provide good assurance that no untoward metabolic, nutritional or antinutritional effects have occurred, an additional and very sensitive measure has been to compare a wide variety of biological characteristics at the whole plant level. The basic question asked is: does the biotech crop fit within the usual definition of that crop? For example, does biotech corn (e.g. BT corn) still possess the expected plant performance of traditional corn? The agronomic and yield characteristics of the genetically modified crop are very sensitive to untoward perturbations in metabolism and in genetic pleiotropy. For example, Table IV lists the parameters that have been assessed for corn. Thus, biotech crops must meet stringent performance criteria.

TABLE III   Compositional Analyses for Corn Grain

Nutrient	YieldGard™ Corn (% dry weight)(16)	Literature Values for Corn (% dry weight)(31, 32)
Protein	13.1	6.0 - 12.0
Fat	3.0	3.1 - 5.7
Fiber	2.6	2.0 - 5.5
Ash	1.6	1.1 - 3.9
Carbohydrate	82.4	na
Calcium	0.0036	0.01 - 0.1
Phosphorus	0.358	0.26 - 0.75·
Amino acid composition	(% total)	(% total)
Alanine	8.2	6.4 - 9.9
Arginine	4.5	2.9 - 5.9
Aspartic acid	7.1	5.8 - 7.2
Cystine	2.0	1.2 - 1.6
Glutamic acid	21.9	12.4 - 19.6
Glycine	3.7	2.6 - 4.7
Histidine	3.1	2.0 - 2.8
Isoleucine	3.7	2.6 - 4.0
Leucine	15.0	7.8 - 15.2
Lysine	2.8	2.0 - 3.8
Methionine	1.7	1.0 - 2.1
Phenylalanine	5.6	2.9 - 5.7
Proline	9.9	6.6 - 10.3
Serine	5.5	4.2 - 5.5
Threonine	3.9	2.9 - 3.9
Tryptophan	0.6	0.5 - 1.2
Tyrosine	4.4	2.9 - 4.7
Valine	4.5	2.1 - 5.2
Fatty acid composition	(% total)	(% total)
Palmitic	10.5	7 - 19
Stearic	1.9	1 - 3
Oleic	23.2	20 - 46
Linoleic	62.6	35 - 70
Linolenic	0.8	0.8 - 2

na = not available

TABLE IV: Biological Equivalence (Corn Example)

Morphological and Agronomic Characteristics	
Stand Establishment	Early plant vigor
Leaf orientation	Leaf colour
Plant height	Root strength (lodging)
Silk date	Silk colour
Ear height	Ear shape
Ear tipfill	Tassel colour
Tassel size	Reaction to fungicides/herbicides
Dropped ears	Late season staygreen/appearance
Stalk rating	Susceptibility to pathogens/pests
Above ear intactness	Yield

Animal Feed Performance Studies

Animal feed performance (nutrition) studies have provided supplementary confirmation of the substantial equivalence and safety of biotech crops. Currently there are many options for animal studies, the choice of which depends on the crop being engineered and its intended use.   Factors evaluated in these studies may include feed intake, body weight, carcass yield, feed conversion, milk yield, milk composition, digestibility and nutrient composition (*33*).

Summary of Food and Feed Safety Assessment

The evaluation of food and feed safety for biotech crops is attacked from two directions, one approaching protein safety and the other approaching whole food and feed safety.   Figure 1 conceptualizes this approach by illustrating the convergence of the risk assessment by considering the weight of all evidence, where no single experiment in the evaluation would be expected to be conclusive alone, as shown in Figure 1.   Using this approach in the evaluation of over 50 biotech crops worldwide, the conclusion has been that foods and feeds derived from biotech crops are as safe and nutritious as foods and feeds derived from traditional crops.

162

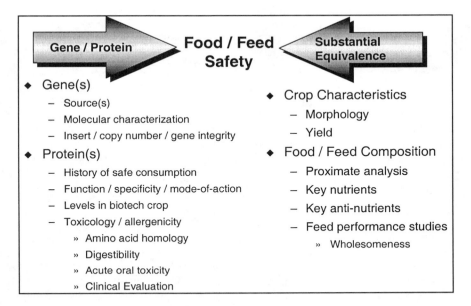

Figure 1 Approach to Food and Feed Safety for Biotech Crops

Conclusions

Adoption of biotech crops by farmers has been driven by tangible economic and productivity gains realized by the agricultural community. More recently, the environmental and food safety benefits of biotech crops have become known, such as the dramatic reduction in pesticide usage in some crops, the encouragement of environmentally sustainable agricultural practices in other biotech crops, and tangible food and feed safety benefits of still other biotech crops. The global scientific community has considered factors important to the food and feed safety of biotech crops, and has concluded that biotech crops should pose no greater risk to animal or human health than traditional foods. Indeed, it has even been said that biotech crops represent the most evaluated foods yet introduced into the food supply. Substantial equivalence and the associated focus on protein safety, compositional and biological analyses of biotech crops have provided the principles upon which food and feed safety assessments are based. The importance of substantial equivalence and its evolution will become even more significant as biotech crops which are intended to provide compositional enhancements and nutritional benefits, such as those created by β-carotene enriched oils, begin to reach the consumer.

163

Acknowledgements

The authors would like to thank S. Astwood, A. Silvanovich and Will Ridley for their helpful comments during the preparation of this manuscript.

References

1. James, C. *Global Review of Commercialized Transgenic Crops: 1999.* 12: Preview, 1-8. 1999. Ithaca, N.Y., The International Service for the Acquisition of Agri-biotech Applications (ISAAA). ISAAA Briefs.
2. Gianessi, L. P. and Carpenter, J. E. *Agricultural Biotechnology: Insect Control Benefits.* 1-65. 1999. Washington, DC, National Center for Food and Agricultural Policy.
3. Munkvold, G. P.; Desjardins, A. E. *Plant Dis.* **1997**, *81*, 556-565.
4. Munkvold, G. P.; Hellmich, R. L.; Rice, L. G. *Plant Dis.* **1999**, *83*, 130-138.
5. Sheehy, R. E.; Kramer, M.; Hiatt, W. R. *Proc. Natl. Acad. Sci.* **1988**, *85*, 8805-8809.
6. Liu, K. *Food Technology* **1999**, *53*, 42-48.
7. WHO. *Malnutrition Worldwide.* 1999. World Health Organization.
8. Gura, T. *Science* **1999**, *285*, 994-995.
9. Shewmaker, C. K.; Sheehy, J. A.; Daley, M.;Coburn, S.; Ke, D. Y. *Plant J.* **1999**, *30*, 1-12.
10. Staub, J. M.; Garcia, B.; Graves, J.; Hajdukiewicz, P. T. J.; Hunter, P.; Nehra, N.; Paradkar, V.; Schlitter, M.; Carrol, J. A.; Spatola, L.; Ward, D.; Ye, G.; Russell, P. F. *Nature Biotech.* **2000**, *18*, 333-338.
11. FAO/WHO. *Biotechnology and Food Safety*; 1996; pp 1-27.
12. FAO/WHO. *Strategies for Assessing the Safety of Foods Produced by Biotechnology*; 1991; pp iii-59.
13. OECD. *Safety Evaluation of Foods Derived by Modern Biotechnology: Concepts and Principles*; Organization for Economic Cooperation and Development (OECD): Paris, 1993.
14. Lavrik, P. B.; Bartnicki, D. E.; Feldman, J.; Hammond, B. G.; Keck, P. J.; Love, S. L.; Naylor, M. W.; Rogan, G. J.; Sims, S. R.; Fuchs, R. L. Safety assessment of potatoes resistant to colorado potato beetle; In *ACS Symposium Series 605. Genetically modified foods. Safety issues.*; Engel, K.-H., Takeoka, G. R., Teranishi, R., eds. American Chemical Society: Washington, 1995; pp 148-158.
15. Padgette, S. R.; Re, D. B.; Barry, G. F.; Eichholtz, D. E.; Delanny, X.; Fuchs, R. L.; Kishore, G. M.; Fraley, R. T. New weed control opportunities: Development of soybeans with a Roundup Ready™ gene; In *Herbicide-*

resistant crops. Agricultural, environmental, economic, regulatory, and technical aspects; Duke, S. O., ed. CRC Lewis Publishers: 1996; pp 53-84.

16. Sanders, P.; Lee, T. C.; Groth, M. E.; Astwood, J. D.; Fuchs, R. L. Safety Assessment of Insect-Protected Corn; In *Biotechnology and Safety Assessment, 2nd ed.*; Thomas, J. A. ed. 1998 Taylor & Francis: 1998; pp 241-256.

17. McClintock, J. T.; Schaffer, C. R.; Sjoblad, R. D. *Pest. Sci.***1995**,*45*, 95-105.

18. Hammond, B. G.; Fuchs, R. L. Safety evaluation for new varieties of food crops developed through biotechnology; In *Biotechnology and Safety Assessment*; Thomas, J. A., ed. Taylor and Francis: Philadelphia, PA, 1998; pp 61-79.

19. Harrison, L. A.; Bailey, M. R.; Naylor, M. W.; Ream, J. E.; Hammond, B. G.; Nida, D. L.; Burnette, B. L.; Nickson, T. E.; Mitsky, T. A.; Taylor, M. L.; Fuchs, R. L.; Padgette, S. R. *J. Nutrition* **1996**, *126*, 728-740.

20. Fuchs, R. L.; Ream, J. E.; Hammond, B. G.; Naylor, M. W.; Leimgruber, R. M.; Berberich, S. A. *Bio/Technology* **1993**, *11*, 1543-1547.

21. Gilissen, L. J. W.; Metz, P. L. J.; Stiekema, W.; Nap, J. P. *Transgenic Research* **1998**, *7*, 157-163.

22. Ewen, P. B.; Pusztai, A. *Lancet* **1999**, *354*, 1353-1354.

23. Frank-Oberaspach, S. L.; Keller, B. *Plant Breeding* **1997**, *116*, 17.

24. U.S. Food and Drug Administration (FDA); Department of Health and Human Services. *Federal Register* **1992**, *57*, 22984-23005.

25. OECD. *Report of The Workshop on the Toxicological and Nutritional Testing of Novel Foods.* SG/ICGB(98)1, 1-48. 9-7-1997.

26. Metcalfe, D. D.; Astwood, J. D.; Townsend, R.; Sampson, H. A.; Taylor, S. L.; Fuchs, R. L. *Crit. Rev. Food Sci. Nutr.* **1996**, *36*, 165-186.

27. Nordlee, J. A.; Taylor, S. L.; Townsend, J. A.; Thomas, L. A.; Bush, R. K. *N. Eng. J. Med.* **1996**, *334*, 688-692.

28. Kimber, I.; Kerkvliet, N. I.; Taylor, S. L.; Astwood, J. D.; Sarlo, K.; Dearman, R. J. *Tox. Sci.* **1999**, *48*, 157-162.

29. Burks, A. W.; Fuchs, R. L. *J. Allergy Clin. Immun.* **1995**, *96*, 1008-1010.

30. Padgette, S. R.; Taylor, N. B.; Nida, D. L.; Bailey, M. R.; MacDonald, J.; Holden, L. R.; Fuchs, R. L. *J. Nutrition.* **1996**, *126*, 702-716.

31. Watson, S. A. Amazing maize. General properties; In *CRC Handbook of Processing and Utilization in Agriculture,* Wolff, I. A., ed. CRC Press: 1982; pp 3-29.

32. Watson, S. A. Structure and composition; In *Corn: Chemistry and Technology*; Watson, S. A., Ransted, P. E., eds. American Asociation of Cereal Chemists, Inc.: 1987; pp 53-82.

33. Hammond, B. G.; Vicini, J. L.; Hartnell, G. F.; Naylor, M. W.; Knight, C. D.; Robinson, E. H.; Fuchs, R. L.; Padgette, S. R. *J. Nutrition* **1996**, 717-727.

Combinatorial Chemistry

When Combinatorial Chemistry emerged in the early eighties, it was initially greeted with skepticism. A good number of synthetic organic chemists were successful in their employment of the classical approach for preparing new compounds. The desired compounds were synthesized one by one and purified using the best methods possible. At first the notion of a "purified mixture" seemed a paradox. However, familiarization breeds comfort, and gradually the advance of the combinatorial approach to synthesis became obvious.

Typically, chemists first attempted the usage of combinatorial reactions with the simplest of all possible techniques. The following is an example of such an approach: A mixture of -- let's say -twenty-five different amines dissolved in an appropriate solvent is reacted with an isocyanate (or an acyl chloride) to produce the corresponding twenty-five substituted ureas (or amides). After evaporating the solvent the mixture was tested for its biological activity. If the mixture showed sufficient activity, the individual compounds were prepared. The combinatorial approach for synthesizing large numbers of compounds within a short period of time greatly improved since its inception with the addition of computer driven machines to do the repetitive chemistry and is speeding discovery of new drugs and agrochemicals

In 1998-1999 we had the delightful experience of organizing a symposium on Combinatorial Chemistry. We have learned to appreciate it as a powerful advance in our field. We would like to take this opportunity to express our gratitude to the Chapter organizers: Dr. B.C. Hamper, Dr. M. Diggelmann, Dr. W.A.Kleschick, Dr. R.D. Gless, Dr. T. Fujita, and Dr. H.Y. Liu, for sharing their great experience and knowledge with us about this important field.

Joseph G. E. Fenyes
1257 N. McLean Blvd.
Buckman Laboratories International, Inc.
Memphis, TN 38108

Isao Iwataki,
Nippon Soda.
Ohtemanti 2-2-1, Chiyoda-ku.
Tokyo 100-8165 JAPAN

Chapter 15

Similarities in Bioanalogous Structural Transformation Patterns

Application to Virtual Library Design for Combinatorial Syntheses of Bioactive Compounds

Toshio Fujita

EMIL Project, #305 Heights Kyogosho, Fuyacho-Nishikikoji-agaru, Nakagyoku, Kyoto 604–8057, Japan
(Formerly, Department of Agricultural Chemistry, Kyoto University)

We have often noticed similarities in structural transformation patterns among various bioactive compound series. Structural transformations of the nitrogen-heterocycle in the methylol moiety of tetramethrin-type insecticides and in protox-inhibiting herbicides are similar to each other in spite of the difference in the pharmacological category. These structural transformations and other observations, if systematized as possible "rules," could be utilizable to design and/or enhance virtual libraries for combinatorial syntheses of bioactive candidates complementarily with the structural diversity.

Structural modification patterns within a series of bioactive analogs are sometimes similar to those observed within another series. This is even the case among series sharing neither a common pharmacology nor a common scaffold structural feature. We note quite a few examples of this type of similarity in structural modification patterns (*1*). One of the typical examples is that among series of Hill reaction inhibitory herbicides, topical antiseptics of cleansing use, and local anesthetics of lidocaine-type. Although the entire structural features are different, the patterns of substructural variations are similar among these series of compounds, occurring in a range of interchangeable amide, urea, carbamate, and related substructural units (*1*).

Another example is among anti-ulcerous H_2-antagonist drugs, insecticidal neonicotinoids, and chromakalim-type anti-hypertensive agents. The substructural components interchangeable within each series are more or less common among these series. These components, so-called polar hydrogen-bonding groups, are such as urea, thiourea, cyanoguanidine, nitroethenediamine, triazolediamine, and related substructural units (*1*).

The interchangeable substructural units in a series of bioactive analogs have been proposed to act as bioanalogous to each other (*1*, *2*). The above examples are thus to show that there are often similarities in bioanalogous (sub)structural transformation/variation patterns among various bioactive compound series regardless of differences in the pharmacological category. In this article, a further example will be presented between a series of insecticidal pyrethroid analogs and a series of herbicidal protox (protoporphyrinogen oxidase) inhibitors. A possible utilitization of this type of information will be discussed in relation to designing virtual libraries in the combinatorial synthesis of an "efficient" discovery of novel candidate lead structures of bioanalogous interest.

Bioanalogous Transformation Examples

Examples in Tetramethrin Analogs

Among synthetic pyrethroids, tetramethrin (**1**) in Figure 1 developed by Sumitomo is noteworthy in its potent knockdown effect on flying insects (*3*). It is widely used as the active ingredient in aerosolic agents against household

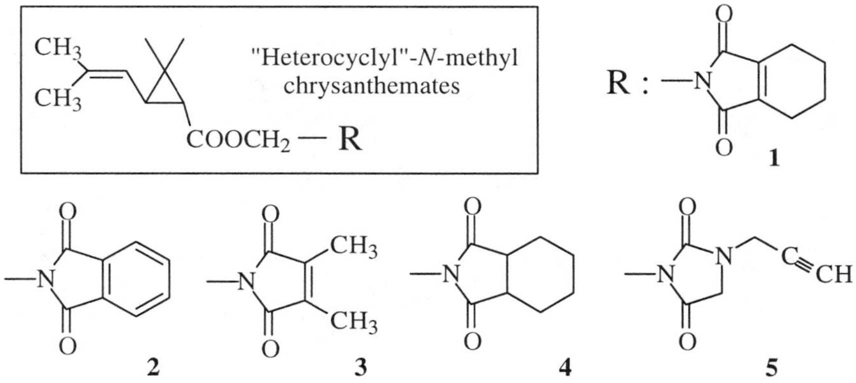

Figure 1. Tetramethrin and its Analogs.

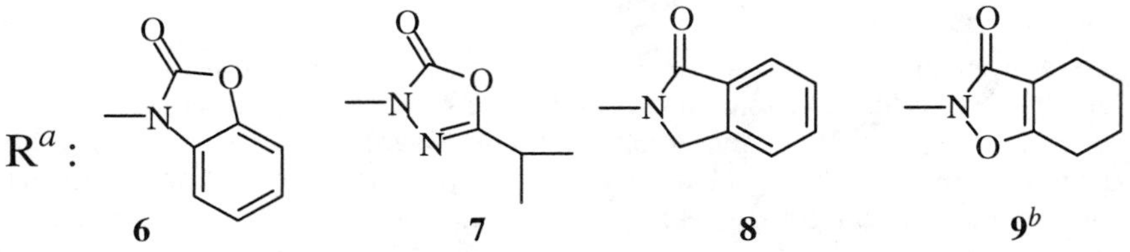

R^a :

6　　　　7　　　　8　　　　9^b

Figure 2. Cyclic amido-N-methylol ester analogs of tetramethrin. (a: Structure of the heterocyclic moiety, see Figure 1. b: The side chain structure of the acid moiety is dichlorovinyl instead of dimethylvinyl in Figure 1.)

pest insects such as houseflies and mosquitoes. The alcoholic moiety is an *N*-heterocyclic dicarboximido-*N*-methylol. Originally, the knockdown and insecticidal activities had been discovered in the corresponding phthalimido derivative (**2**) at Sumitomo (*4*). Following these compounds, a number of analogs with various *N*-heterocyclic *N*-methylols have been synthesized. Compounds (**3,4**) in Figure 1 are also knockdown-active (*4,5*). Imiprothrin (**5**), disclosed by Sumitomo, is more knockdown-active than tetramethrin (*6*). It has been utilized as an aerosolic ingredient against cockroaches.

Structural modifications into mono-carbonyl heterocyclic *N*-methylol esters (**6-9**) shown in Figure 2 are found in the literature (*5,7-9*). They are also knockdown-active in various degrees.

Further modifications into tri-nitrogen-heterocyclic analogs (**10-13**) in Figure 3 had been conducted at Sumitomo (*10,11*). The *N*-substituent of the heterocycles is propargyl similar to that in imiprothrin (**5**). They are mostly more knockdown-active than tetramethrin (**1**), but none has been commercialized. Some compounds in Figures 2 and 3 possess the dichlorovinyl in place of the dimethylvinyl side chain in the acid moiety.

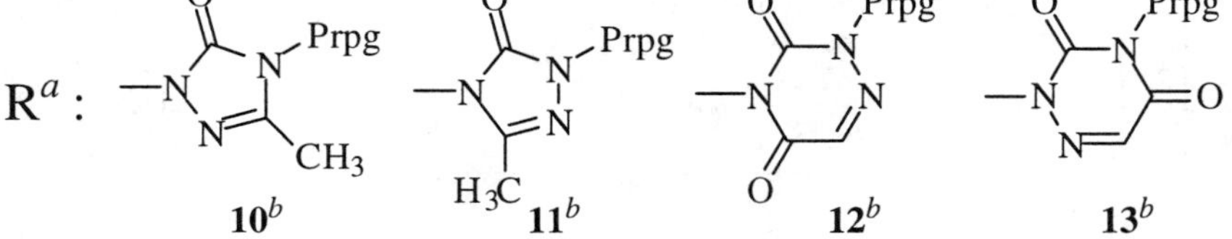

R^a :

10^b　　　　11^b　　　　12^b　　　　13^b

Figure 3. Tri-nitrogen-hetrocyclic analogs of tetramethrin. (a: See note a of Figure 2; Prpg = propargyl, -CH$_2$C≡CH. b: See note b of Figure 2.)

The pyrrole- and pyrazole-*N*-methylol esters (**14, 15**) shown in Figure 4 have also been synthesized, indicating that they are knockdown-active, if appropriate substituents are placed onto the heterocyclic ring (*12*). The exocyclic C=O double bonds in imido- and amido-*N*-methylols in Figures 1-3 seem to be replacable with the endocyclic C=N as well as C=C double bonds in these compounds.

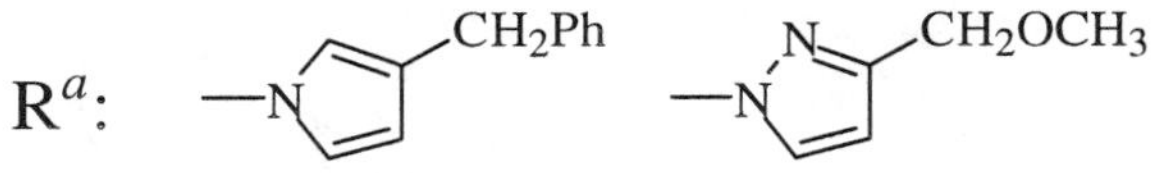

R^a:

14[b] **15**[b]

Figure 4. Pyrrole- and pyrazole-N-methylol esters.
(a: See note a of Figure 2; b: See note b of Figure 2.)

Examples in Protox-Inhibiting Herbicides

The earliest members of *N*-phenyl-nitrogen-heterocyclic protox inhibitors are oxadiazon (**16**) developed by Rhone-Poulenc (*13*) and chlorophthalim (**17**) by Mitsubishi (*14*) shown in Figure 5. When they were first introduced as herbicides of agricultural use, their mode of action such as the protox inhibition had been unknown. Requiring light conditions for their activity, they had been called as light-dependent herbicides for years. Structural collation between oxadiazon (**16**) and chlorophthalim (**17**) had also been unclear. The substitution patterns in the phenyl moiety of chlorophthalim have been extensively explored leading to flumipropyn (**18**) and flumiclorac-pentyl (**19**), both developed by Sumitomo (*15,16*), and many others. The structures of the cyclic "amido" component of oxadiazon (**16**) and "imido" moiety of chlorophthalim (**17**) have also been modified variously to give compounds such as pentoxazone (**20**) of Kaken (*17*), sulfentrazone (**21**) and carfentrazone-ethyl (**22**) of FMC (*18,19*), and those (**23-27**) from other organizations (*20-24*) shown in Figure 5. Thus, the two earliest compounds are recognized as possessing common structural features: an appropriately substituted (mostly 2,4,5-trisubstituted) phenyl group and a pertinent nitrogen-heterocycle. Substituents attached to the phenyl moiety are mostly F at the 2-(*ortho*) and Cl at the 4-(*para*) position, while the substituent at the 5-position is more complicated. Typically, it is either a modified alkoxy, including propargyloxy and alkoxycarbonylalkoxy, or a substituted sulfonamido group. Substituents on the heterocycles are, however, difficult to be generalized. Note that nipyraclofen (**27**) developed by Bayer has a 2,4,6-tri-substitution pattern on the phenyl moiety similar to fipronil-type insecticides, the general structure of which also belongs to the *N*-phenyl nitrogen-heterocycles (*25*).

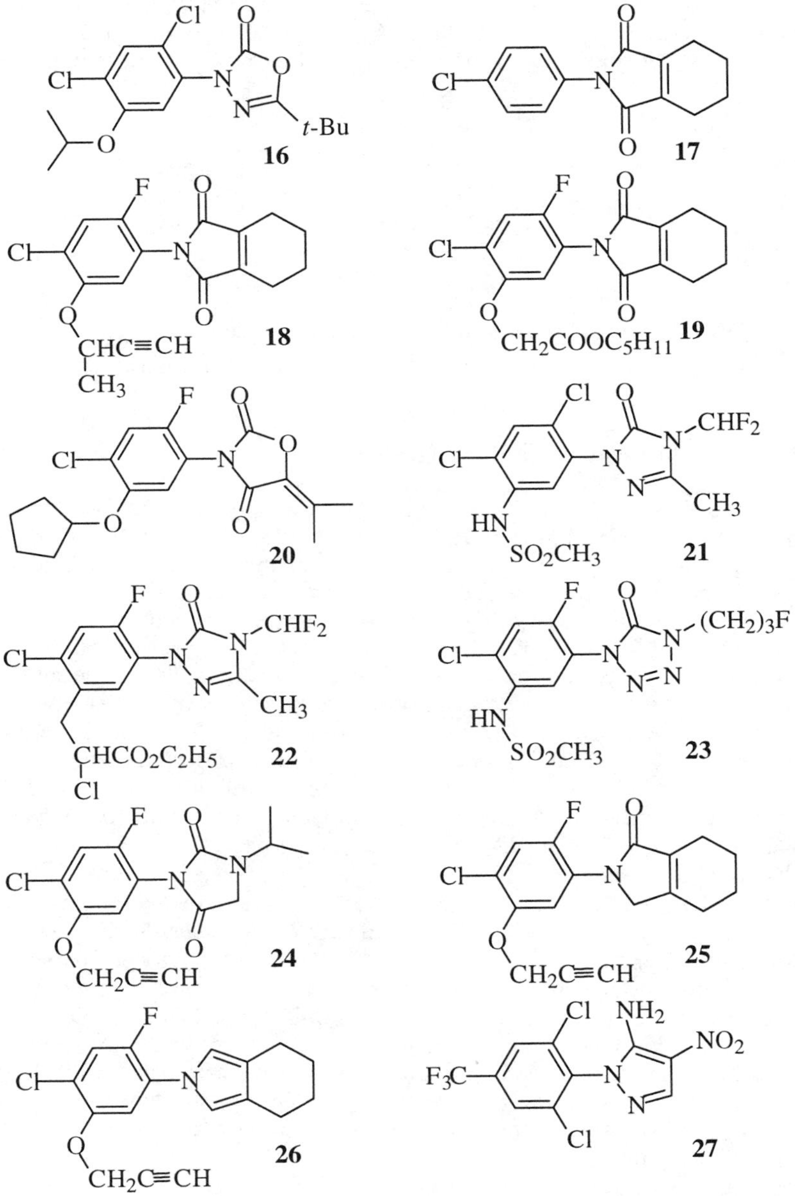

Figure 5. *Oxadiazon, chlorophthalim, and their analogs with 5-membered (condensed) N-heterocycles.*

Various 6-membered ring analogs (**28-33**) are shown in Figure 6 (*26-31*). The number of nitrogen atoms within the heterocycle seems to be allowed at least from unity to three. As observed between compounds **16** and **21**, the ether oxygen and alkylated nitrogen are replaceable between compounds **28** and **29**. Endocyclic N=C bonds in compounds (**30-32**) can be regarded as the exocyclic C=O in compounds (**28, 29**) having been incorporated within the ring.

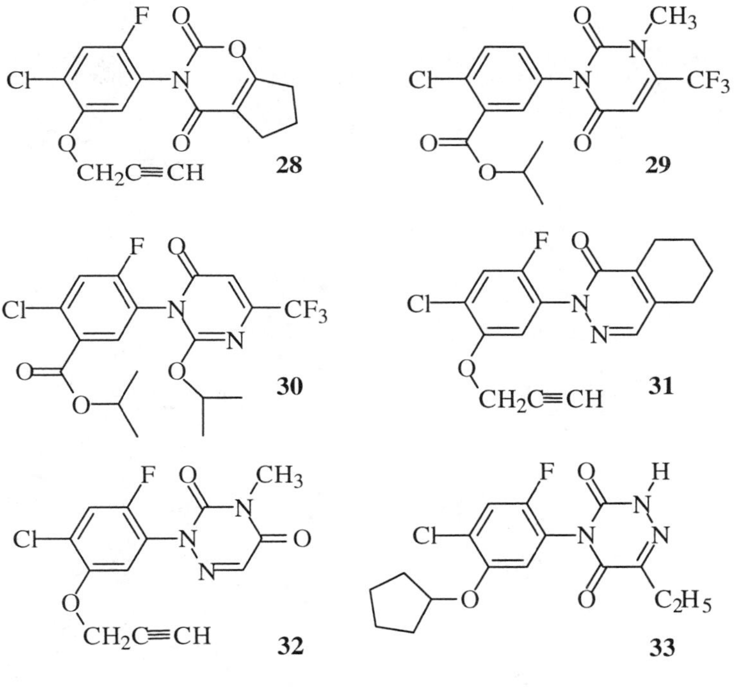

Figure 6. 6-Membered N-heterocyclic analogs of oxadiazon/chlorophthalim.

There are a number of bioanalogous structures/compounds in protox inhibiting herbicides other than those described above. To cover structures more comprehensively, an excellent review recently published should be consulted (*32*).

172

Similarity in Structural Transformation/Variation Patterns

Bioanalogous Substructure Library "Common" Among Various Series

As can be easily recognized, there are many heterocyclic substructures common between the above two series of compounds, although the substituents are often considerably different (*33*). The situation is summarized in Figure 7, where the flows of (sub)structural transformations of heterocyclic moiety such as those from exocyclic C=O to endocyclic C=C and C=N double bonds, the replacement of ether –O– with –NR–, and the expansion of 5-membered to 6-membered ring systems are arranged somewhat systematically, according to the possible "evolution" routes. Each of the arrows indicates that the structural "evolution" could be achieved stepwise. At the same time, the heterocyclic structures as a set can be regarded as a bioanalogous substructure library.

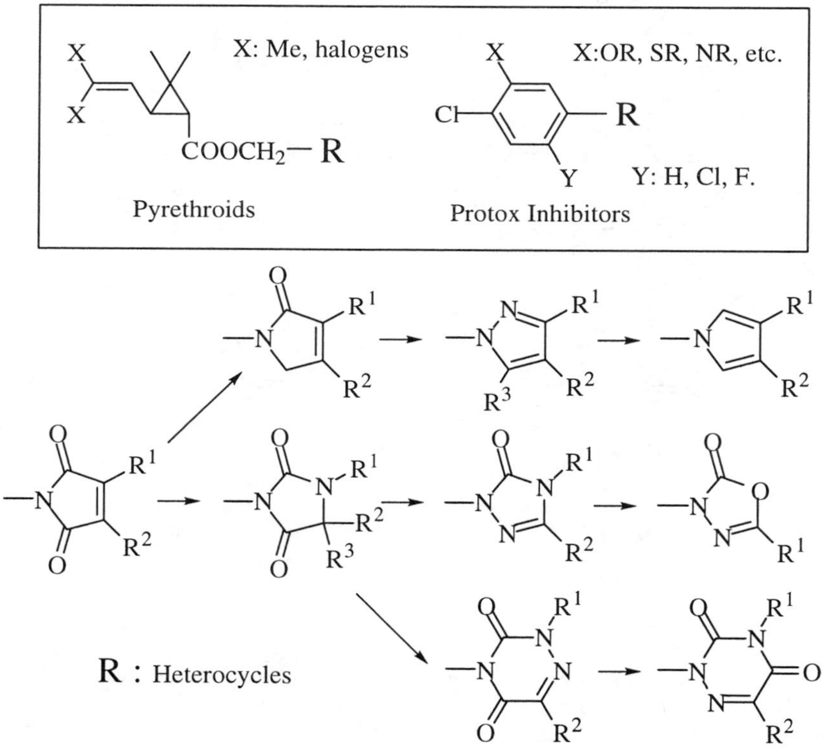

Figure 7. Library of bioanalogous heterocyclic structures shared by tetramethrin-type pyrethroids and protox-inhibiting herbicides.

When various structural moieties or substructural units are interchangeable within a series of bioactive compounds without losing the activity, these substructural units have usually been called **bioisosteric**. Because the term **bioisosterism** inherently involves the concept of isosterism or isometricity, Floersheim *et al.* (*34*) first proposed replacing the term with **bioanalogy** which could apply to cases in which the structural dimensions are not (very) close but more or less drastically varied. Among heterocyclic ring systems alone shown in Figure 7, it could be stated that there is an approximate bioisosterism. The size of 5- and 6-membered rings is, however, not exactly equivalent. Including substituents that have been optimized for each of the particular compounds, the term bioanalogy should be understood to be more reasonable. Such a similarity in bioanalogous substructural transformation patterns as that shown in Figure 7 has also been found in non-acidic antiinflammatory agents of the pyrine-type as shown in Figure 8 (*33*) as well as in fipronil series insecticides (not shown) in a range somewhat narrower.

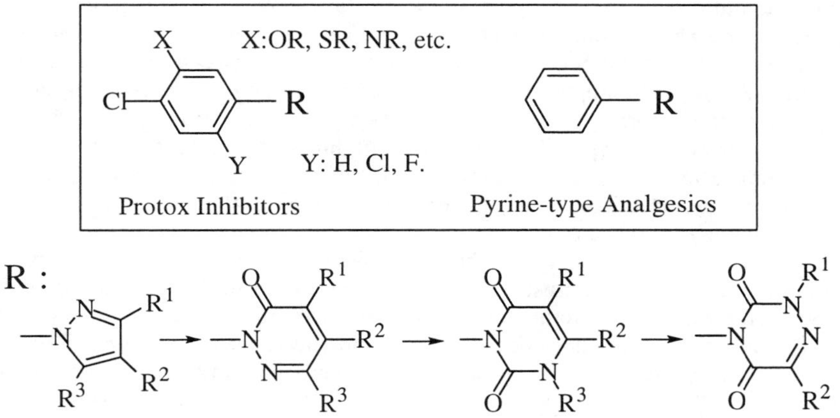

Figure 8. Library of bioanalogous heterocycles shared by protox inhibiting herbicides and pyrine-type analgesics.

Structural Transformation Patterns for a Database of a System "EMIL"

With the interchangeability as a tool, it could be possible to "design" the structural modification routes leading to "novel" bioanalogous structures, whenever a possible prototype structure shares the maleimido moiety or one of the substructure members in the library shown as Figure 7. The structural modification could be done either stepwise by replacing the initial moiety with other members consecutively or batchwise simultaneously to every member substructure in the library. The modification procedures could apply naturally to series of pyrethroid alcohols, protox inhibiting herbicides, pyrine analgesics, and fipronil-type insecticides. Further, it could be extended to other bioactive compound series of different pharmacology as well. It should be noted, however, that bioactivity of the virtual compounds is unknown at this stage. In Figures 7 and 8, substituents attached to the ring systems are just symbolized. In transformed compounds in which the ring system has been replaced by one of the library members and certain bioactivities have indeed been disclosed, substituent(s) and the substitution patterns should be variably optimized depending upon disclosed bioactivities.

As mentioned in the introductory section, there are quite a few precedents in which similarities in bioanalogous (sub)structural transformation patterns exist among various series of compounds regardless of the differences in the pharmacological category. Thus, various bioanalogous (sub)structures in each series are potentially able to recognize the micro-environmental features of the corresponding receptor site in a manner similar to each other. Among various proteins, the amino acid sequence is by no means random, but some of the sequence segments are believed to be more or less exactly conserved regardless of differences in the apparent function of proteins (35). If certain conserved segments or similar segments are located at the interaction site of various receptor proteins responsible for different bioactivities, the bioanalogous substructure members could interact with "various" receptor segments in a similar manner regardless of the origin as well as the apparent function (pharmacology) of receptors.

It is thus believed that precedents of bioanalogous (sub)structural transformation patterns can almost be regarded as a set of "rules" which could guide how to forward the structure evolution from candidate lead structures. The pharmacological differences could not be concerned so seriously in each of the evolution processes in the first place. We have been working on a project to integrate and organize these precedents as a database. If the database is incorporated into a system in which any initially input structures are processed with the "rules" to release elaborated higher-ordered structures as the output "automatically", the system could be a great benefit for the synthetic chemists of bioactive compounds. We have published quite a few (review) articles in which such a system named EMIL (**E**xample-**M**ediated-**I**nnovation-for-**L**ead-Evolution) is interpreted and discussed (1,2,36,37). In this system, bioanalogous structural transformation precedents published in the past as well as just newly disclosed or patented should be collected as comprehensively as possible. For detail of the EMIL system, previous articles should be consulted.

The EMIL System and the Library Design for Combinatorial Synthesis

In combinatorial synthesis of bioactive compounds, the library design is one of the most important aspects. An example of the conventional procedure would be simplified as follows. A "discovery" library, with which to start the high-throughput screening, may include a large number of compounds. Their origin may be from a corporate library commercially available and/or a compound collection stored for years in synthetic chemical institutions. To avoid redundancy, the library members are selected so that they are as diverse as possible in multidimensional structure-based parameter space (diversity selection) (38-40). "Hits" are identified as primary lead compounds by the screening for particular bioactivities. Then, structures of the hits are searched for whether or not they share a common pharmacophoric framework, i.e., whether or not a certain structure-activity pattern could be hypothesized among them, either 2- or 3-dimensionally (structure-activity hypothesis). The structures, which are conformable to the pharmacophore hypothesis, are designed with the aid of computer algorithms. When necessary, the virtual

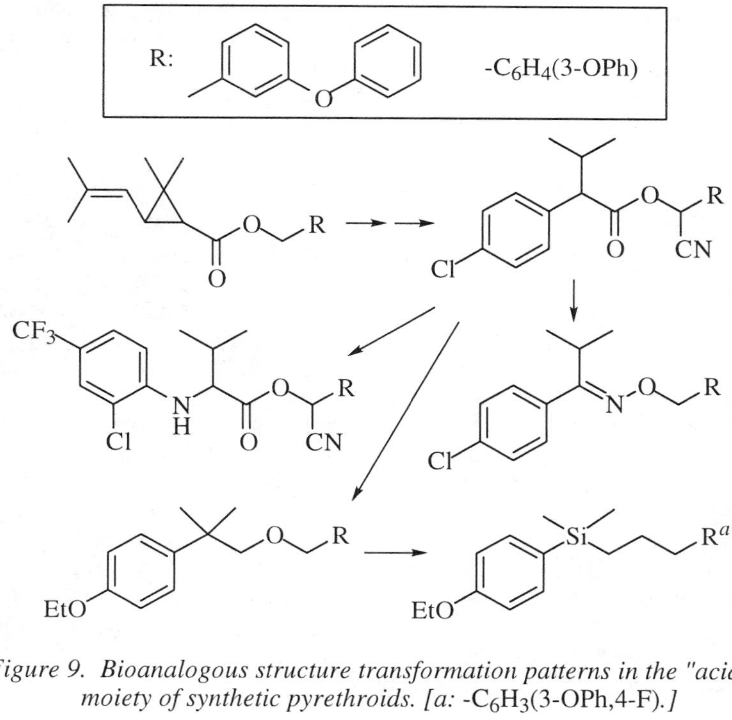

Figure 9. Bioanalogous structure transformation patterns in the "acidic" moiety of synthetic pyrethroids. [a: -C$_6$H$_3$(3-OPh,4-F).]

compounds that are consistent with the pharmacophore hypothesis are added to the primary hits to construct a virtual library (virtual library design) (*41*). After elaboration of the virtual library by deleting some compounds in terms of the difficulty of synthesis and inappropriate physicochemical properties computationally (*42*), (parallel) synthesis of the library members, the high-throughput screening, and the pharmacophore search are repeated to detect the higher-ordered lead structures/compounds, consecutively.

The most significant aspect in the application of the EMIL system to the combinatorial synthesis would be to help designing the virtual library. It could be enhanced by incorporating novel members transformed batchwise from the original hits' structures with the aid of bioanalogous substructure transformation "rules". The bioanalogous structures are inevitably "analogous". Differing from bioisosteric structures which could more or less follow the concept of the isosterism/isometricity, however, such drastic structural modification patterns as observed in the acidic moiety of pyrethroid insecticides shown in Figure 9 (*37*) are also defined as being bioanalogous and stored in the database. In fact, the substructural transformation "rules" in the EMIL database are classified and scored in terms of the "drasticity" of the transformation. By selecting the patterns (rules) of higher scores for transformations towards geometrically dissimilar structures, considerable diversity could still be kept in the virtual library. On the other hand, patterns of lower scores connecting geometrically "similar" bioanalogous substructures such as *N*-heterocycles shown above could be utilized to design "bioanalog-directed/biased" libraries (*41*), efficiently (*43*).

Conclusions

To date, theoretically defined and/or computationally derived criteria and/or descriptors, *e.g.*, those based on the "structure-based design" procedures including the "docking study", have been used for the virtual library design in combinatorial synthesis (*44*). The present procedure according to an empirical concept with bioanalogous (sub)structural transformation patterns as a potential descriptor has never been introduced into the library design technology.

In the past structural transformation series of bioactive compounds such as pharmaceuticals and agrochemicals, a majority of individual steps may originally be attempted on trial-and-error basis. This is also true with the combinatorial synthesis. Because structural transformation patterns included in these steps have eventually been "utilized" in improving or at least in retaining the bioactive profile successfully, they are well regarded as being invaluable information for the analog design or bioanalogous molecular transformation. The present type of an empirical knowledge-based system is expected to be of a practical use also in the combinatorial library design, perhaps, complementarily with theoretical/computational algorithms.

Acknowledgments

The author would like to thank Drs. Nobuyuki Okajima, Central Pharmaceutical Research Institute of Japan Tobacco Inc., Yoshihisa Inoue, Research Institute of Yoshitomi Pharmaceutical Co., and Ferenc Darvas, ComGenex, Inc. for their invaluable suggestions, and Dr. Yoshiaki Nakagawa, Section of Bioregulation Chemistry of Kyoto University for his skillful artwork. Drs. Yoshihiro Minamite of Dainippon Jochugiku Company, Ltd., Kenji Hirai of Sagami Chemical Research Center, Ko Wakabayashi of Tamagawa University, and Chiyozo Takayama of Sumitomo Chemical Company are greatly appreciated for their kind help collecting literature.

References

1. Fujita, T. *Biosci. Biotech. Biochem.* **1996**, *60*, 557-566.
2. Fujita, T.; Adachi, M.; Akamatsu, M.; Asao, M.; Fukami, H.; Inoue, Y.; Iwataki, I.; Kido, M.; Koga, H.; Kobayashi, T.; Kumita,I.; Makino, K.; Oda, K.; Ogino, A.; Ohta, M.; Sakamoto, F.; Sekiya, T.; Shimizu, R.; Takayama, C.; Tada, Y.; Ueda, I.; Umeda, Y.; Yamakawa, M.; Yamaura, Y.; Yoshioka, H.; Yoshida, M.; Yoshimoto, M.; Wakabayashi, K. In *QSAR and Drug Design : New Developments and Applications* ; Fujita, T., Ed.; Pharmacochemistry Library 23 ; Elsevier : Amsterdam, **1995**; pp 235-273.
3. Kato, T.; Ueda, K.; Fujimoto, K. *Agr. Biol. Chem.* **1964**, *28*, 914-915.
4. Yoshioka, H. In *Rational Approaches to Structure, Activity, and Ecotoxicology of Agrochemicals*; Draber, W.; Fujita, T., Eds.; CRC Press: Boca Raton, FL, **1992**; pp 185-217.
5. Nishimura, K.; Kitahaba, T.; Ikemoto, Y.; Fujita, T. *Pestic. Biochem. Physiol.* **1988**, *31*, 155-165.
6. Hirano, M.; Itaya, N.; Ohno, I.; Fujita, Y.; Yoshioka, H. *Pestic. Sci.* **1979**, *10*, 291-294.
7. Blinova, V. G.; Ivanova, S. N.; Shvetsov-Shilovskii, N. I.; Mel'nikov, N. N. U.S.S.R. Patent SU 305763, **1968**.
8. Ueda, K.; Mizutani, T.; Yoshioka, H.; Fujimoto, K.; Okuno, Y. Japanese Patent JP 44021331B4, **1969**.
9. Ohno, I.; Takeda, H.; Nishioka, T.; Itaya, N. Jpn. Kokai Tokkyo Koho, JP 53092768, **1978**.
10. Kishida, H.; Yano, T. Jpn. Kokai Tokkyo Koho, JP 57158765A, **1982**.
11. Mizutani, M.; Tsushima, K.; Sanemitsu, Y.; Hirano, M. European Patent EP 33163A2, **1981**.
12. Ohsumi, T.; Hirano, M.; Itaya, N.; Fujita, Y. *Pestic. Sci.* **1981**, *12*, 53-58.
13. Boesch, R.; Metivier, J. European Patent EP 957151, **1963**.

14. Matsui, K.; Kasugai, H.; Matsuya, K.; Aizawa, H. Japanese Patent JP 48011940, **1973**.
15. Nagano, E.; Hashimoto, S.; Yoshida, R.; Matsumoto, H.; Oshio, H.; Kamoshita, K. European Patent EP 61741A2, **1982**.
16. Nagano, E.; Hashimoto, S.; Yoshida, R.; Matsumoto, H.; Kamoshita, K. European Patent EP 83055A2, **1983**.
17. Hirai, K.; Futikami, T.; Murata, A.; Hirose, H.; Yokota, M.; Nagato, S. WO 8702357A1, **1987**.
18. Theodoridis, G.; Baum, J. S.; Hotzman, F. W.; Manfredi, M. C.; Maravetz, L. L.; Lyga, J. W.; Tymonko, J. M.; Wilson, K. R.; Poss, K. M.; Wyle, M. J. In *Synthesis and Chemistry of Agrochemicals III*; Baker D. R.; Fenyes, J. G.; Steffens, J. J., Eds.; ACS Symposium Series 504; American Chemical Society : Washington, DC, **1992**; pp 134-146.
19. Theodoridis, G.; Bahr, J. T.; Dividson, B. L.; Hart, S. E.; Hotzman, F. W.; Poss, K. M.; Tutt, S. F. In *Synthesis and Chmistry of Agrochemicals IV*; Baker, D. R.; Fenyes, J. G.; Basarak, G. S., Eds.; ACS Symposium Series 584; American Chemical Society : Washington, DC, **1995**; pp 90-99.
20. Theodoridis, G. WO 8703873A1, **1987**.
21. Nagano, E.; Hashimoto, S.; Yoshida, R.; Matsumoto, H.; Kamoshita, K. Jpn. Kokai Tokkyo Koho, JP 58219167A2, **1983**.
22. Schallner, O.; Negele, M.; Santel, H.-J.; Luerssen, K.; Schmidt, R. R.; Krauskopf, B. U.S. Patent 5024694, **1991**.
23. Kume, T.; Goto, T.; Kamochi, A.; Yanagi, A.; Miyauchi, H.; Asami, T. European Patent EP 351641A1, **1990**.
24. Schallner, O.; Gehring, R.; Klauke, E.; Stetter, J.; Wroblowsky, H.-J.; Schmidt, R. R.; Sante, H.-J. U.S. Patent 4614533, **1986**.
25. Cole, L. M.; Nicholson, R. A.; Casida, J. E. *Pestic. Biochem. Physiol.* **1993**, *46*, 47-54.
26. Sante, H.-J.; Kunisch, F.; Ooms, P.; Strang, H.; Schmidt, R. R. U.S. Patent 5061310, **1991**.
27. Bell, A. R. U.S. Patent 4943309A, **1990**.
28. Wengler, J.; Winternitz, P. European Patent EP 260621A2, **1988**.
29. Saito, K.; Ito, E.; Ishikawa, H.; Inaba, H.; Sato, J. Jpn. Kokai Tokkyo Koho, JP 63156779, **1988**.
30. Lyga, J. W.; Halling, B. P.; Witkowski, D. A.; Patera, R. M.; Seeley, J. A.; Plummer, M. L.; Hotzman, F. W. In *Synthesis and Chemistry of Agrochemicals II* ; Baker, D. R.; Fenyes, J. G.; Moberg, W. K., Eds.; ACS Symposium Series 443; American Chemical Society : Washington, DC, **1991**; pp 170-181.
31. Hirai, K.; Shikakura, K.; Yano, T.; Ishikawa, C.; Ugai, S.; Yamada, O. WO 9622285A1, **1996**.
32. Hirai, K. In *Peroxidizing Herbicides*; Böger, P.; Wakabayashi, K., Eds.; Springer: Berlin, **1999**, pp 15-71.

33. Fujita, T.; Nakayama, A. In *Peroxidizing Herbicides*; Böger, P.; Wakabayashi, K.; Eds.; Springer : Berlin, **1999**, pp 91-139.
34. Floersheim, P.; Pombo-Villar, E.; Shapiro, G. *Chimia*, **1992**, *46*, 323-334.
35. Cornish-Bowden, M. *Nature*, **1985**, *313*, 434-435.
36. Fujita, T. In *Classical and Three-Dimensional QSAR in Agrochemistry;* Hansch, C.; Fujita, T.; Eds.; ACS Symposium Series 606; American Chemical Society : Washington, DC, **1995**; pp 13-34.
37. Fujita, T.; Nishimura, K.; Cheng, Z.-M.; Yoshioka, H.; Minamite, Y.; Katsuda, Y. In *Natural and Engineered Pest Management Agents*; Hedin, P. A.; Menn, J. J.; Hollingworth, R. M., Eds.; ACS Symposium Series 551; American Chemical Society : Washington, DC, **1994**; pp 396-406.
38. Ghose, A. K.; Viswanadhan, V. N.; Wendoloski, J. J. *J. Comb. Chem.* **1999**, *1*, 55-68.
39. McGregor, M. J.; Muskal, S. M. *J. Chem. Inf. Comput. Sci.* **2000**, *40*, 117-125.
40. Rusinko, A., III; Farmen, M. W.; Lambert, C. G.; Brown, P. L.; Young, S. S. *J. Chem. Inf. Comput. Sci.* **1999**, *39*, 1017-1026.
41. Haque, T. S.; Skillman, A. G.; Lee, C. E.; Habashita, H.; Gluzman, I. Y.; Ewing, T. J. A.; Goldberg, D. E.; Kuntz, I. D.; Ellman, J. A. *J. Med. Chem.* **1999**, *42*, 1428-1440.
42. Lipinski, C. A.; Lombardo, F.; Dominy, B.W.; Feeney, P. J. *Adv. Drug Delivery Rev.* **1997**, *23*, 3-25.
43. A Windows version of the EMIL system suited for the library design is developed by ComGenex, Inc., So. San Francisco, CA, U.S.A. and Budapest, Hungary (Postal address: P.O. Box 73, Budapest 62, H-1388, Hungary, http://www.comgenex.com & http://www.comgenex.hu).
44. Walters, W. P.; Stahl, M. T.; Murcko, M. A. *Drug Discovery Today* **1998**, *3*, 160-173.

Chapter 16

Preparation of Small Molecule Libraries for Agrochemical Discovery

Development of Parallel Solution and Solid Phase Synthesis Methods

Bruce C. Hamper, Thomas J. Owen, Kevin D. Jerome, Angela S. Scates, Stephen A. Kolodziej, Robert C. Chott, and Allen S. Kesselring

Monsanto Company, AG Sector, 800 North Lindbergh Boulevard, St. Louis, MO 63167

Using a combination of solid phase and solution phase techniques, we have developed methods for the streamlined preparation of targeted compounds for our discovery efforts. Acrylate, malonic acid and bromoacetate resins have been used for the preparation of oligomeric β-amino acids, pyrimidinediones, methylene substituted malonic acid derivatives and trifluoromethylketo (TFMK) imidazoles. A PlateView system has been developed to facilitate the archival and analysis of analytical data (HPLC, GC, LCMS, NMR, etc.) associated with individual compounds from these libraries. Using combinatorial analytical chemistry tools, sets or libraries of compounds can be evaluated for structural confirmation and purity. The tools developed for visualization of multiple chromatographic or NMR spectral data will be described using examples from the Monsanto AG Sector chemical discovery effort.

We have been involved in the preparation of compounds using the tools of combinatorial chemistry and parallel synthesis methods for agrochemical discovery for most of the 1990's (*1-5*). As is true of many discovery organizations, the advent of high throughput screens has enabled us to test the entire set of historical compounds in our collection against new targets in a very short period of time. Parallel synthesis methods have allowed us to augment or supplement our historical collection with compounds specifically designed for investigation of activity against specific targets. Significant gains in productivity can be achieved using parallel versus sequential processing for preparation of new compounds. Methods which allow efficient parallel processing, whether solid phase or solution phase, have become the cornerstone of combinatorial chemistry (*6,7*). Arguably there is nothing new about parallel processing and one of the greatest practitioners of modern times was Thomas Edison. One can observe in his laboratory museum in Orange, New Jersey an extensive setup of parallel extractors for obtaining latex from goldenrod plant samples. In his last studies, Edison built a large steam table capable of handling over two dozen Soxlet extractors in an attempt to find a commercially useful source of latex or natural rubber (*8*). Parallel processing allowed him to determine the relation of varieties of goldenrod with latex content in an efficient manner. As agricultural scientists, we can only wonder what would have been possible if Edison had the tools of genetic engineering at his disposal for enhancing latex content.

Combinatorial chemistry encompasses more than just the gains in efficiency of parallel processing and for a given library provides a means of preparing compounds or products encompassing all possible combinations of a set of inputs or reagents (*9*). In order to employ combinatorial chemistry efficiently for the preparation of compound libraries, one needs 1) efficient solution and/or solid phase synthesis methods and 2) a means of tracking each compound and its associated physical or biological data. Laboratory automation and reactor design can be very beneficial in the handling of various aspects of chemical library production. However, even with minimal expenditures in equipment, the tools of combinatorial chemistry can be employed successfully. This is particularly true for the preparation of small libraries which are often used in targeted synthesis programs. At the same time that our department was setting up the infrastructure for combinatorial chemistry, one of our projects involved preparation of a series of herbicidal phenylpyrazoles which are inhibitors of protoporphorinogen oxidase (*10*). These studies led to the identification of commercial candidate JV 485, a useful pre-emergent herbicide which provides season-long, broad-spectrum control of weeds in winter wheat (*11*). Parallel synthesis methods allowed the rapid preparation of an important set of 4-alkylpyrazole derivatives which could be compared directly for herbicidal activity in whole plant screens (Figure 1).

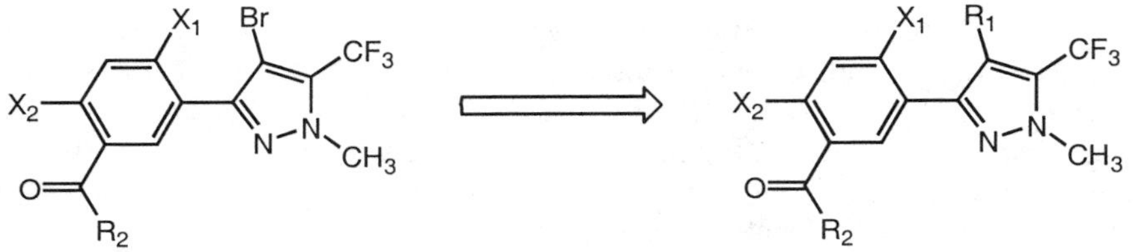

1
3-Aryl-4-halopyrazoles
JV 485 (X_1 = F, X_2 = Cl, R_2 = O-*i*-propyl)

2
Analogs prepared by parallel synthesis
R_1 - (halo)alkyl; R_2 - OR, NHR
US 5,668,088 (1997)

Figure 1. Parallel synthesis of herbicidal analogs of JV485.

Since combinatorial methods were used to prepare this series, all possible combinations of the 4-alkylpyrazole (R_1) and the 5'-phenyl (R_2) substituents were available for direct comparison of the structure-activity relationships (*12*). Following this initial successful use of parallel synthesis methods for investigation of compounds relevant to agricultural chemical discovery, we pursued development of a wider range of synthetic targets. In this chapter we will present the method development and preparation of three libraries as examples of this implementation; oligomeric β-peptoids, malonate derivatives, and trifluoromethylketo (TFMK) imidazoles.

Oligomeric *N*-Substituted-β-Alanines (β-Peptoids)

In an effort to develop more efficient methods for parallel synthesis, we investigated the use of novel solid phase intermediates for the generation of larger libraries of compounds. The first of these intermediates was acrylate **3** (a derivative of Wang resin), which was used for the preparation of pyrimidinediones **5** (*4*) and β-peptoids **6** (*13*) (Figure 2). Acrylate resin **3** can be prepared in quantitative yield by treatment of Wang's resin with acryloyl chloride (Figure 3). Treatment of the acrylate resin with excess amine in DMSO affords resin bound *N*-substituted-β-alanines **4**. Although the Michael addition is a reversible reaction, complete conversion is obtained by using an excess of amine. One of the initial difficulties we faced was determination of the loading of the Wang-acrylate resin and the Michael addition product. Although the conversion can be monitored qualitatively by FTIR of individual beads, the actual loading can not be directly measured using this technique. In principle, the acrylic acid obtained on cleavage of the resin with TFA can be measured, however the volatility of the cleaved material precludes concentration of the sample and determination of the product by gravimetric analysis. We developed a "Direct Cleavage NMR" method in which the resin loading can be determined by ^{1}H NMR measurement of the cleaved filtrate in the presence of an internal standard (13,14).

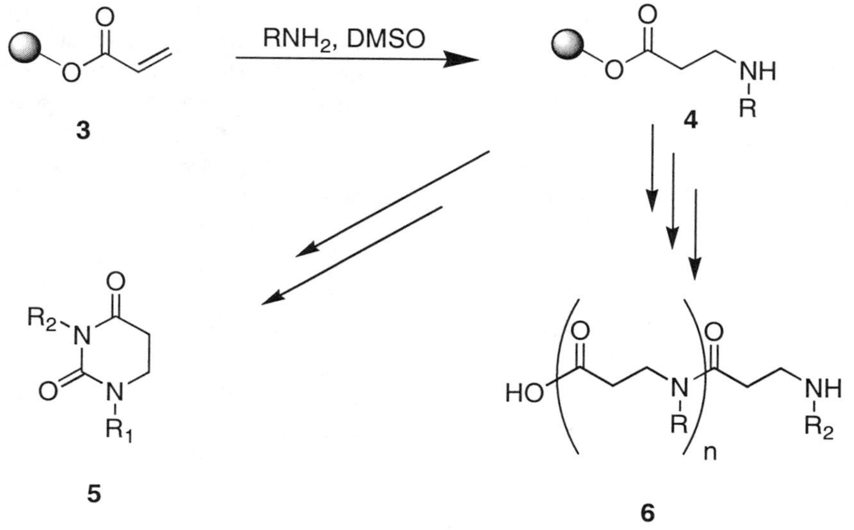

Figure 2. Use of acylate resin 3 for the preparation of pyrimidinediones 5 and oligomeric β-peptoids.

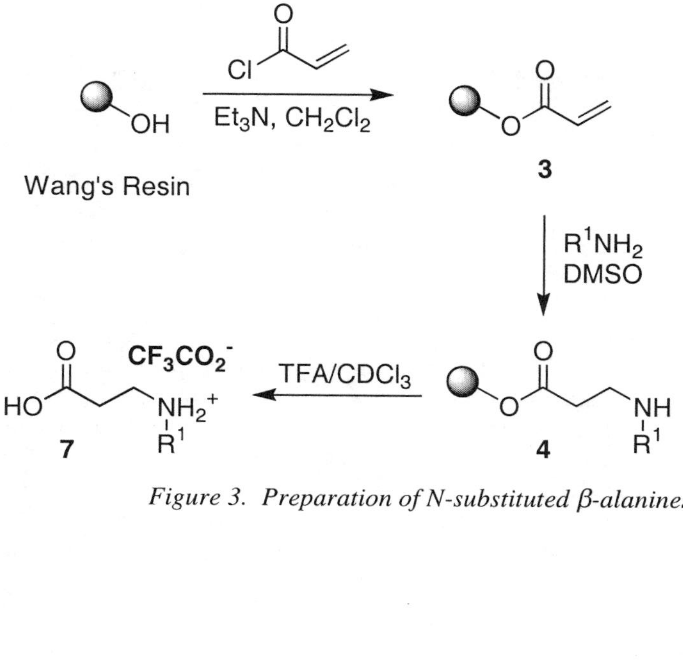

Figure 3. Preparation of N-substituted β-alanines 7.

The resin cleavage is carried out using NMR compatible solvents containing a measured amount of an internal standard. Direct cleavage of a weighed amount of resin with a 10 mM solution of hexamethyldisiloxane (HMDS) in CDCl₃/TFA (1:1) provided an NMR compatible sample without concentration of the filtrate solution. This method has the advantage of detecting impurities which may be present in the resin and have not been adequately removed by washing, such as residual solvents or reagents. Direct cleavage NMR of the *N*-benzyl-β-alanine resin **4a** (R^1 = benzyl) shows only the cleaved β-amino acid **7a** without any measurable amount of acrylic acid starting material (Figure 4). Loading of the product resin **4a** can be determined by integration of the product resonances and the HMDS internal standard to give a calculated loading of 0.79 milliequivalents per gram.

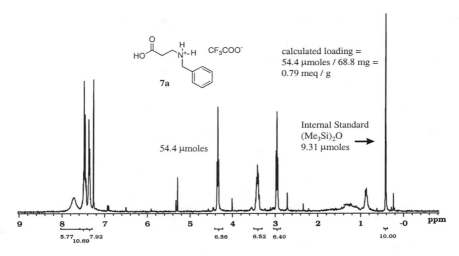

Figure 4. Direct Cleavage ^{1}H NMR of N-benzyl-β-alanine 7a and determination of resin loading of 4a. The integration values of the three methylene resonances (2.96, 3.42 and 4.35 ppm) were averaged and used to calculate the amount of 4a present in the NMR solution.

The Michael addition of amines to acrylates for preparation of *N*-substituted-β-alanines has also been investigated by solution phase, however, mixtures of mono- and bis-addition products are obtained along with unreacted amine (*15*). The separation problems that ensue have not made this an attractive preparative method. By comparison, the solid phase route is preparatively useful and provides the desired products in pure form. This is a case in which, rather than adapt well known chemistry to solid phase, we have taken advantage of the attributes of resin based chemistry (allowing the use of excess reagents to drive a reaction to completion and providing for easy removal of reagents by filtration) to develop chemistry uniquely suited to solid phase.

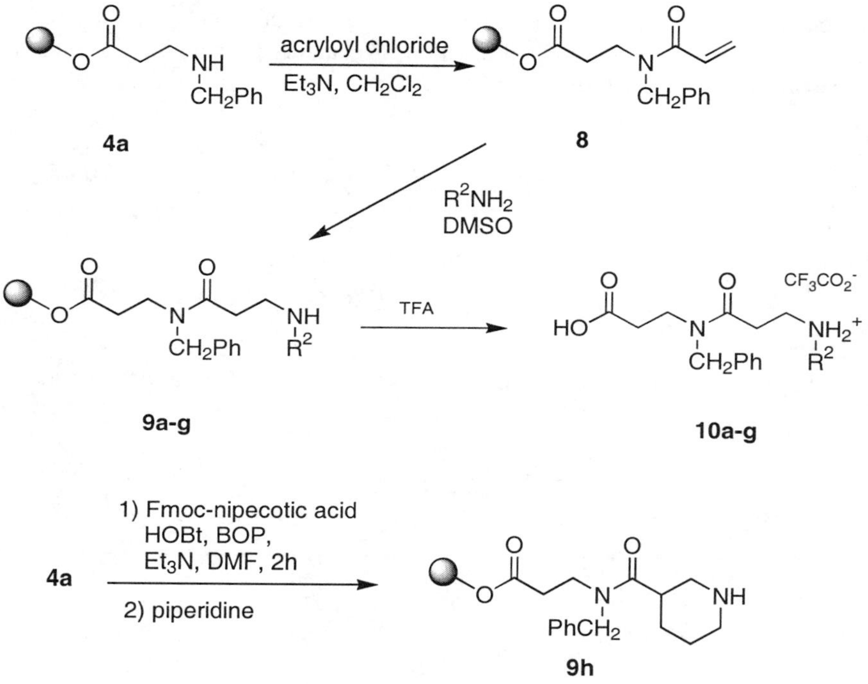

Figure 5. Preparation of di-β-peptoids by solid phase synthesis.

Once the monomeric β-amino acids **4** are obtained on a solid phase, sequential addition of acylate and amines can provide the basis for construction of dimeric N-substituted β-amino acids or di-β-peptoids (Figure 5). The second β-amino acid is added to **4a** by a two step sequence via acylation with acryloyl chloride to give acrylamide **8** followed by Michael addition of an amine to give dimer **9**. Not surprisingly, addition of amines to the acrylamide requires a higher temperature than the first addition to acrylate **3**, due to the lower reactivity of acrylamides to nucleophiles. Cleavage of **9** affords the dimers **10** in overall yield for the two steps from **8** of 70-92% with conversions in excess of 95% for most examples (Table 1). Hindered amines such as cyclopropylamine or ethyl alanate (entries e,f) afforded lower conversions in the comparative 24 h reactions, however these cases can be driven to completion by double treatments with amine as is common in many solid phase reactions. Less nucleophilic substrates, however, such as anilines and *t*-butylamine fail to react with the acrylamide resin **8** even under longer or repeated treatments. As would be expected, standard amino acid coupling procedures can be used to incorporate Fmoc protected amines. This technique was used not only for the addition of α-amino acids, but also for incorporation of unique β-amino acids such as nipecotic acid as in the formation of dimer **9h**.

Table 1. Preparation of Di-β-peptoids (10) From Acrylamide (8) and Amines.

Entry	R-NH₂	Yield (%)ᵃ	Conversion (%)ᵃ	HPLC (min., area %)ᵇ
a	benzyl-NH₂	85.2	>95	2.09 (93%)
b	iso-butyl-NH₂	92.5	>95	1.67 (91%)
c	sec-butyl-NH₂	77.9	>95	1.50 (94%)
d	iso-propyl-NH₂	81.7	>95	1.10 (91%)
e	cyclopropyl-NH₂	73.2	90	1.03 (86%)
f	EtOOCCH₂CH₂-NH₂	72.8	83	0.98 (85%)
g	phenyl-NH₂	0	0	0

ᵃYield and conversion were determined by direct cleavage ^{1}H NMR. The yield represents the percent μmoles of product compared to theoretical. Conversions were determined by comparison of the acrylamide and product **10** resonances. ᵇHPLC retention times and area percent of major peak.

The combined techniques of sequential acylation/Michael addition or standard peptide couplings were used to prepare a set of trimeric β-peptoids **12** by multistep solid phase synthesis from Wang acrylate resin **3** (Figure 6). A total of 16 amines/amino acids were chosen as inputs for trimer formation providing a library of 4096 (16 x 16 x 16)

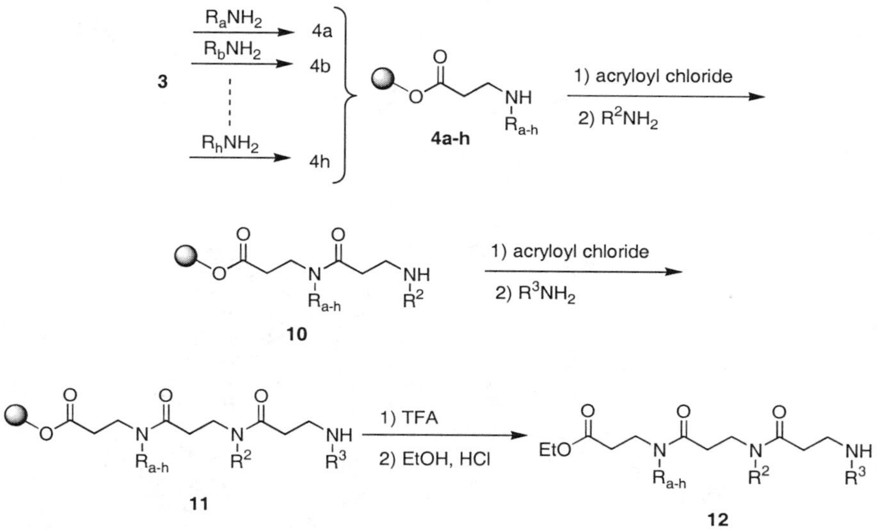

Figure 6. Preparation of library of trimeric β-peptoids 12 as their N-acetyl ethyl ester derivatives.

compounds. Eventually, this synthesis was expanded to provide many more examples, however the first set is instructive for discussion of the approach to the library. Depending on the available time and automation capability, it is possible to prepare this set as individual samples with each compound isolated in an individual well or reaction vessel. However, from a production standpoint, it is much more efficient to prepare the compounds as mixtures. At the time, we were interested in the β-peptoids as lead generation libraries which can be prepared for screening in a mixture format consisting of eight individual compounds per reaction well. As active leads were identified, the individual members of a mixture were re-synthesized for evaluation and confirmation in the screens. Using a split and mix technique, the number of compounds per well and their relative concentrations can be carefully controlled. Initially, eight resins **4a** through **4h** were prepared in eight separate reactors and analyzed individually to determine the molarity of each resin. An equimolar mixture of the eight resins **4a-h** was prepared and divided into eight vessels for addition of the second β-amino acid to give eight new resin mixtures **10**. The third amino acid addition required 64 individual reaction vessels to obtain the final products **12**. A parallel synthesis apparatus or STAR block was devised to carry out parallel solid phase reactions and allow washing of the resins prior to cleavage (*16*). All of the resin bound trimers **11** were cleaved to provide the tri-β-peptoids as free acids. Esterification of the complete library gave 512 mixtures containing 8 products each for a total of 4096 trimeric β-peptoid ethyl esters **12**.

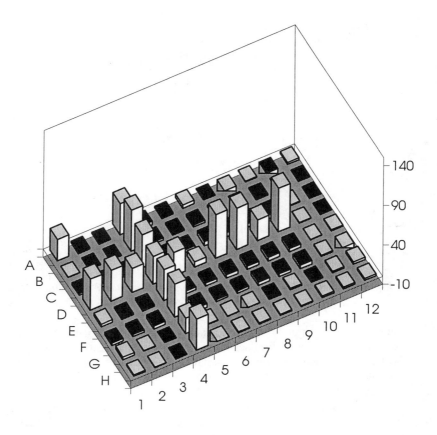

Figure 7. Fungicide summary data for a set of 96 mixture samples containing eight compounds per well.

Fungicidal activity was found for a number of the mixture samples from the oligomeric β-amino acid library and corresponded to particular substituents in the R2 and R3 positions of the β-peptoids. In a typical set of 96 samples, each sample well contains a mixture of eight components corresponding to different substituents in the R1 position of **12**. The rows A-H and the columns 1-12 designate the R2 and R3 substituents, respectively. Therefore, trends in activity for particular substituents can be determined directly from a graphical representation of activity for a plate of 96 samples. Fungicidal summary data for one such plate (Figure 7) shows activity for row D and column 4, both of which correspond to *n*-dodecylamine in the R2 and R3 positions. The cross peak at D4 did not give an enhancement in activity for this set. In order to evaluate the effect of the dodecyl group, we prepared single compounds for testing for each of the presumed components of the pooled samples in row D and column 4. Only limited biological activity was obtained with the deconvoluted samples, indicating that the combined activity of eight components in the original mixture was required for activity. This is a case in which the individual components have similar, albeit weak, fungicidal activity and produce an additive effect in combination in the original screening of pooled samples.

Knoevenagel Condensation: Preparation of Unsymmetrical Malonate Intermediates

Early in the development of combinatorial chemistry methods, we realized that the availability of unique intermediates on solid phase resins can facilitate the preparation of libraries. 1,3-Dicarbonyl compounds and their equivalents (e.g., 1,3-diketones, β-ketoesters, malonates, α,β-unsaturated esters, etc.) are very useful for the preparation of heterocycles by condensation with bidentate nucleophiles such as hydrazine or hydroxylamine. After investigation of a number of possible routes, we found that malonic acid resin **13** (Figure 8) can be readily prepared by treatment of Wangs resin with Meldrum's acid (*17*). Unsymmetrical malonic acid **13** provides simultaneous protection of one of the carboxylic acid groups via attachment to the resin and the ability to derivatize the free acid. Cleavage of these materials from the Wang resin gives unsymmetrical amide or ester derivatives of malonic acid.

Knoevenagel condensations were investigated using a set of eight malonic acid resins **14{A-H}** prepared from **13** and eight alcohols or amines (R^1OH or R^1NH_2) (*18*). The eight resins were placed in individual vessels of a 96 reaction STAR block such that **14{A}** was in the 12 vessels in row A, **14{B}** in row B, etc. Once the 96 vessels were filled with the appropriate resin, the vessels were treated with aldehydes to give Knoevenagel adducts **15{A-H, 2-12}**. The resins in column 1 were left untreated to provide standards for unreacted malonic acid derivatives after cleavage from the resin. The vessels in columns 2 through 12 were treated with different aldehydes such that a total of 88 methylene substituted malonic acids **16** were obtained. HPLC analysis of

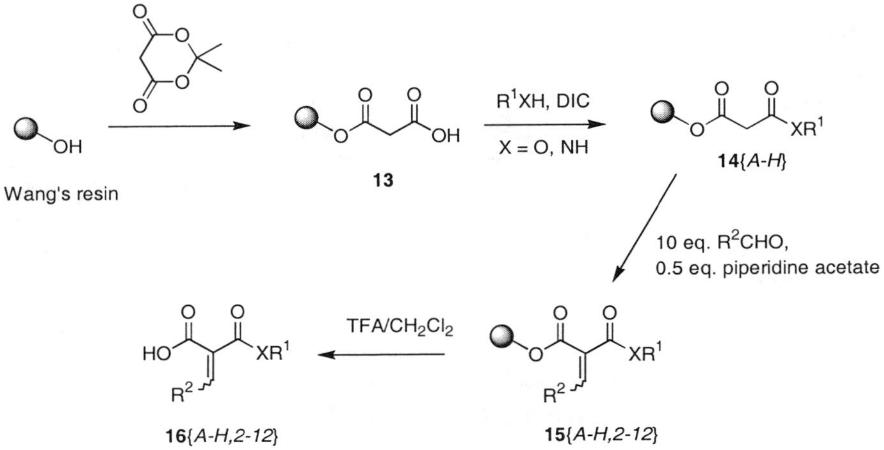

*Figure 8. Preparation of methylene substituted malonamic acids and esters **16** from malonic acid resin **13**.*

the cleavage products show good conversion of the starting materials (column 1) to the expected products for the majority of the reactions as seen in the PlateView diagram (Figure 9). A clear cell indicates less than 20% starting material left in the final product, whereas a shaded cell indicates that the reaction failed to go to completion and contains greater than 20% starting material. This set of 96 compounds was also investigated by NMR using a flow-through probe VAST system. The hardware for these systems has been developed primarily for LC-NMR investigations, however by connecting the probe to a liquid handling robot capable of transferring solutions from deep-well microtiter plates, the samples from small to moderate size compound libraries can be introduced into the probe and returned to the microtiter plate for archival. Analysis of the ^{1}H NMR spectra obtained for this set of 96 compounds revealed trends which were nearly identical to that obtained by HPLC. In both cases the reactions in column 6 did not provide the expected products with the exception of vessel A6. The correlation between these two methods suggests that medium to high-throughput 1H NMR will be very useful in the future for analysis of libraries.

A - PlateView of HPLC data - Percent of Starting Material

	1	2	3	4	5	6	7	8	9	10	11	12
A	100	0	0	0	0	0	10	5	0	15	5	0
B	100	0	0	0	0	100	0	0	0	0	0	0
C	100	19	0	0	7	94	6	5	0	6	0	0
D	100	0	0	0	0	90	0	0	0	0	0	0
E	100	7	3	0	4	49	0	0	0	2	2	2
F	100	14	11	4	12	100	11	8	3	3	5	5
G	100	13	8	5	16	24	8	10	5	8	0	5
H	100	35	22	22	20	100	18	16	12	21	0	22

B. PlateView of ^{1}H NMR Data - Confirmation of Structure

	1	2	3	4	5	6	7	8	9	10	11	12
A	100	100	50	100	100	50	50	100	100	50	100	100
B	100	100	100	100	100	0	50	100	50	50	50	50
C	100	50	50	50	50	0	50	100	100	50	100	100
D	100	50	50	50	50	0	100	100	100	100	100	100
E	100	100	100	100	100	50	100	100	100	100	100	100
F	100	50	50	100	50	0	100	100	100	100	100	100
G	100	50	100	50	50	50	100	50	50	50	100	100
H	100	50	50	100	100	0	100	100	100	100	100	100

Figure 9. PlateView analysis of HPLC (A) and NMR (B) results from 96 component malonamic acid library 16. The clear cells indicate a successful reaction, whereas the shaded cells indicate that the product contains greater than 20% starting material (HPLC) or does not contain the desired product (as determined by NMR).

Solid Phase/Solution Phase Approach: TFMK-Imidazoles

One of the more interesting developments in discovery organizations in the last few years has been the use of parallel synthesis methods for preparation of arrays of compounds specifically targeted towards particular targets. These lead optimization studies require knowledge of yield and purity of individual members in order to obtain structure-activity relationships from the biological data. The desired final compounds do not always lend themselves to the high yield parallel synthesis methods which have been developed to date. As a result, the preparation of these targeted sets can be quite challenging! We faced just such a case in the preparation of imidazole trifluoromethylketones (TFMK-imidazoles) (Figure 10).

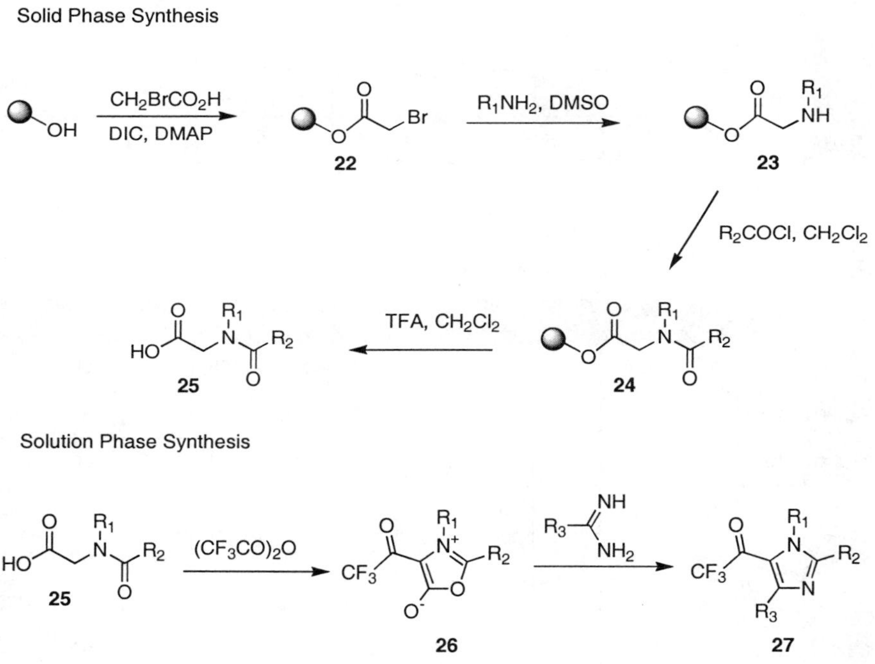

Solid Phase Synthesis

Solution Phase Synthesis

Figure 10. Preparation of TFMK imidazoles 27 by a combination of solid and solution phase chemistry.

Compounds containing the TFMK group, particularly peptide derivatives, have been shown to be potent inhibitors of a variety of proteases and esterases. Based on modelling sudies, we were interested in the potential of TFMK-imidazoles to inhibit carboxyl terminal D1 processing protease (CtpA). This protease is critical for the production of D1 protein in plants, which is the target of photosystem II (PSII) herbicides such as atrazine and diuron (*19, 20*).

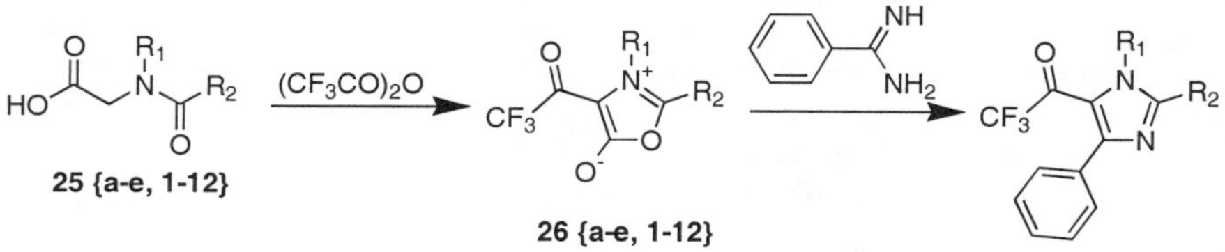

	1	2	3	4	5	6	7	8	9	10	11	12
A	97	97	97	94	34	97	89	96	94	97	66	58
B	57	88	96	98	4		93	73	99	96	21	85
C	59	84	70	93		12	85	93	98	90	97	81
D	80	92	97	98		79	70	96	56	82	85	99
E	74			85	31	69			51	67	42	65

Figure 11. PlateView analysis of 4-phenyl-5-trifluoromethylketoimidazoles 27. The clear cells indicate greater than 50% yield of the desired product, whereas the shaded cells indicate less than 50% product. Cells without a number indicate that no desired product was obtained.

The TFMK-imidazoles **27** were prepared by a mixed solid/solution phase motif to allow introduction of three different R groups in the ring system (*21*). Very few N-substituted glycines are commercially available, so a solid phase route was used which allows independent introduction of an amine, followed by an acylating group to give glycine derivatives **25**. A set of 60 glycines **25** was prepared and analyzed by LC/MS. Parallel solution phase methods were used to obtain the desired TFMK-imidazoles. The unique set of 60 glycines was treated in parallel with trifluoroacetic anhydride to give munchnones **26** which were

194

treated with an amidine to provide the imidazoles **27**{a-e, 1-12, A}. Yields obtained from the solution phase dipolar cycloaddition varied greatly as seen in a sample set of 60 compounds prepared from benzamidine (Figure 11). However, a majority of the compounds were obtained in greater than 80% yield. In seven of the cases (B6, C5, D5, E2-3, E7-8) no measurable product was obtained. Prior to submission to a biological assay, compounds below the desired purity level were purified using an automated preparative HPLC system. PlateView diagrams of LC/MS or nmr results were critical for evaluation of sets of samples and determining which samples need further processing or purification prior to submission to an assay. A number of the TFMK-imidazoles showed evidence of activity against CtpA in the low micromolar range and also showed activity *in vivo* against arabidopsis in whole plant screens.

Conclusions

Over the last ten years, we have seen the development of combinatorial chemistry techniques from the first initial efforts of specialized groups to the adaptation of these techniques for routine use in chemical discovery. It is now fair to state the parallel synthesis techniques are in common practice in the industry and can be employed either for initial lead discovery from large diverse sets of compounds or used for rapid targeted lead development after identification of an active area of chemistry. We expect to see many more applications using a targeted lead development approach in concert with molecular modeling, rational drug design and traditional single component synthesis. As we enter the new millennium, the challenges of the next decade will be determination of the quality and integrity of chemical libraries. Advances have been made in terms of rapid collection and analysis of LC/MS data and will be followed by similar developments in NMR, LC-NMR and other coupled techniques. This will provide the quality control required for the use of combinatorial libraries in the evaluation of structure-activity relationships in biological systems.

References

1. Hamper, B. C.; Dukesherer, D. R.; South M. S. *Tetrahedron Lett.* **1996**, *37*, 3671-3674.
2. Hamper, B. C.; Dukesherer, D.; South, M. S. *212th American Chemical Society National Meeting Book of Abstracts*, August 25-29, 1996, ORGN-188, *American Chemical Society*: Washington, D.C., 1996.
3. Hamper, B. C.; Kolodziej, S. A. *212th American Chemical Society National*

Meeting Book of Abstracts, August 25-29, 1996, ORGN-082, *American Chemical Society*: Washington, D.C., 1996.

4. Kolodziej, S. A.; Hamper, B. C. *Tetrahedron Lett.* **1996**, *37*, 5277-5280.
5. Parlow, J. J.; Mischke, D. A.; Woodard, S. S. *J. Org. Chem.* **1997**, *62*, 5908-5919.
6. Dolle, R. E.; Nelson, K. H. *J. Combinatorial Chem.* **1999**, *1*, 235-282.
7. *Combinatorial Chemistry and Molecular Diversity in Drug Discovery*; Gordon, E. M.; Kerwin, J. F.; Eds.; Wiley-Liss: New York, NY, 1998.
8. Vanderbilt, B. M. *Thomas Edison, Chemist*; American Chemical Society: Washington, D. C., 1971, pp. 273.
9. Wilson, S. R. In *Combinatorial Chemistry*; Wilson, S. R.; Czarnik, A. W.; Eds.; John Wiley & Sons: New York, NY, 1997; pp 1-24.
10. Hamper, B. C.; Mao, M. K.; Phillips, W. G. U.S. Patent 5,698,708, 1997.
11. Prosch, S. D.; Ciha, A. J.; Grogna, R.; Hamper, B. C.; Feucht, D.; Dreist, M. *Brighton Crop Prot. Conf.-Weeds*; 1997, Vol. 1, pp. 45-50.
12. Hamper, B. C.; McDermott, L. L. U.S. Patent 5,668,088, 1997.
13. Hamper, B. C.; Kolodziej, S. A.; Scates, A. M.; Smith, R. G.; Cortez, E. *J. Org. Chem.* **1998**, *63*, 708-718.
14. Hamper, B. C.; Kesselring, A. S. In *Solid-Phase Organic Synthesis*, Czarnik, A. W.; Ed., in press.
15. Stork, G.; McElvain, S. M. *J. Am. Chem. Soc.* **1947**, *69*, 971-972.
16. Hamper, B. C. U.S. Patent 5,792,430, 1998.
17. Hamper, B. C.; Kolodziej, S. A.; Scates, A. M. *Tetrahedron Lett.* **1998**, *39*, 2047-2050.
18. Hamper, B. C.; Snyderman, D. M.; Owen, T. J.; Scates, A. M.; Owsley, D. C.; Kesselring, A. S.; Chott, R. C. *J. Comb. Chem.* **1999**, *1*, 140-150.
19. Duff, S. M. G.; Fabbri, B. J.; Yalamanchili, G.; Remsen, E. E.; Michener, M. L.; Hamper, B. C.; Walker, D. M.; CaJacob, C. A. *Plant Physiol.* **1999**, *120*, 132-S ASPP Conference Abstract #599.
20. Oelmueller, R.; Herrmann, R. G.; Pakrasi, H. B. *J. Biol. Chem.* **1996**, *271*, 21848-21852.
21. Hamper, B. C.; Jerome, K. D.; Yalamanchili, G.; Walker, D. M.; Chott, R. C.; Mischke, D. A. *Biotech. Bioeng. (Polymer Supports for Streamlined Organic Synthesis)*, in press.

Chapter 17

Parallel Synthesis at Novartis Crop Protection: Concept and Realization

M. Diggelmann, J. Ehrler, and W. Lutz

Novartis Crop Protection AG, Lead Discovery Department,
P.O. Box, 4002 Basel, Switzerland

In order to increase synthesis capacity and to support lead finding and lead optimization activities, sizable investments into laboratory automation and state-of-the-art purification equipment were made at Novartis Crop Protection. Successful implementation of a *high-speed-synthesis* platform consisted of the modular assembly of different commercially available workstations following an off-line automation approach in combination with the design of cheap and primitive custom made reaction manifolds. Major emphasis was always given to the objective of optimizing overall throughput which led early to the simple conclusion that the bottleneck of isolating pure compounds is of utmost importance. The increased ability to derive reliable structure-activity relationships, fundamental for the tasks of chemists, should consequently lead to a fast return of capital investments spent to address this issue.

Combinatorial chemistry has emerged as a powerful tool for lead generation as well as lead optimization, and it's importance is demonstrated by the size of investments and the speed of implementation made by all major pharmaceutical and agrochemical companies. The clear driving force is the desire for faster drug discovery cycles (*1*). Since it's invention about a decade ago (*2,3*), a number of different techniques and ingenious concepts were described in the literature, ranging from truly combinatorial synthesis on solid support (*4,5*) towards more automated parallel synthesis approaches favoring liquid phase (*6,7,8*). While initially people were impressed by the huge number of compounds that can be prepared by these techniques, it also became quite clear over the years that simply preparing and screening more compounds does not automatically solve the drug optimization problem, among others probably one of the reasons why currently much focus is on improving the cheminformatics environment in our industry (*9,10*). Challenges and difficulties in finding marketable drugs or pesticides obviously still exist, and this will lead according to our view to a much healthier situation with respect to a realistic assessment of the potential of combinatorial chemistry. That such a potential certainly exists nobody seriously denies engaged in the art of drug design.

Impact on Drug Discovery

The large number of existing combinatorial techniques provides today's researchers with a new set of tools, the selection of which largely depends on the characteristics of the problem to be solved. It's probably fair to say that the majority of applications in <u>solid-phase</u> synthesis are used in a true combinatorial fashion and applied more towards lead finding while <u>solution-phase</u> based approaches, supported by variable degrees of laboratory robotics, seem better geared for lead optimization or hit-follow up activities as indicated in Figure 1.

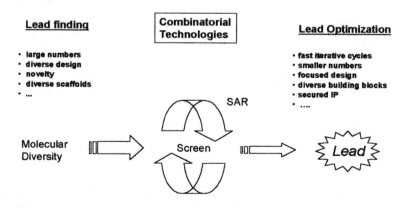

Figure 1: Impact on drug discovery by combinatorial technologies

Automation Concepts

At Novartis Crop Protection solid-phase as well as liquid-phase based approaches were and still are being pursued, this article only illustrating our activities in the latter one. Over the last years improvements in laboratory robotics and data management contributed significantly to the rapid growth of parallel synthesis methods. Conceptually two approaches can and should be distinguished. One which we termed on-line automation, meaning that several different tasks are performed by the same device, and if implemented successfully, allows long run times of the involved machine(s) without any manual intervention. The second one, off-line automation, refers to the modular integration of several different workstations into an assembly line, each performing a dedicated task. A clear distinction between these two automation approaches is more than just semantics since they differ significantly in overall performance and investment costs. Reasons for us to favor off-line automation are outlined in Table I.

Table I: Comparison of Automation Approaches

Issue	On-Line	Off-Line
• # of reactors in parallel	small	high (scalable)
• adaption to technical improvements	difficult	easy
• removal of bottlenecks	impossible	possible
• user interface	challenging	easy
• throughput	low	high
• manual steps	no	yes (> racks)

As becomes clear from studying the table, the realization of a *high throughput synthesis* environment by an off-line automation concept based on liquid-phase synthesis is easy, low tech and cheap. The reason for this statement is the fact that the synthesis and isolation processes involved (Figure 2) are the same as the ones performed traditionally by chemists for decades.

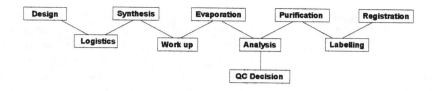

Figure 2: Workflow in Parallel Synthesis

Parallel liquid-phase synthesis can therefore be performed with rather simple devices (Table II), and all that's left to do is basically to design reactors that allow increased parallel handling. For overall efficiency of course consideration about batch size and interfaces between the different devices are important and crucial for success, e.g. assigning a standard rack format is a clear must. In addition one has to ask oneself carefully what, how and how much to automate.

Table II: Devices for parallel liquid-phase Synthesis

Process	Device
1. Adding BB's and Reagents	Pipette Robot
2. Heating	Hotplate
3. Quenching	Pipette Robot
4. Liquid Extraction	Pipette Robot
5. Evaporation	Evaporator
6. Dissolving	Pipette Robot
7. Aliquot Pipetting for Analysis	Pipette Robot
8. Evaporation	Evaporator
9. Weighing	Balance
10. Dissolve	Pipette Robot
11. Aliquot Pipetting for Screening	Pipette Robot

Reactor Design

At the time of conception, no really cheap reactors were commercially available, and we decided to work with our own racks custom made from aluminum. Objectives guiding the design were overall low costs, flexible use and the ability to be used in connection with already existing commercially available liquid handling robots. We favored a 6x4 format similar to the one used by biologists for cell culturing and also relying on the same footprint as the traditional 96-well microtiter plates. This design gave us the flexibility to work in reasonable batch sizes, as well as the opportunity to work at different synthesis scales; and, additionally, e.g. the ability to perform liquid-liquid extraction steps directly out of the synthesis reactor. Depending on the reaction conditions to be applied, either disposable PP based tubes or reusable glass reactors can be used. Heating and stirring is performed by traditional magnetic hot plate stirrers and magnetic stirrer bars while parallel sealing of the reactor device is achieved via mechanical clamping, analogous to the principle described many years ago by Meyers et al. (11). A schematic representation of the reactor manifold is shown in Figure 3.

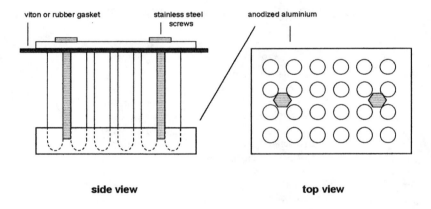

Figure 3: Reactor Design

Feasible Chemistry

In these admittedly rather simple reactor blocks, a wide range of reactions are possible allowing the chemist, in combination with simple liquid handlers and temperature controlled liquid cooling bathes, even to perform quite delicate organometalic chemistry (Figure 4). This may seem at first glance surprising, but more careful thinking about the physical prerequisites for such reactions leads to the conclusion that such experiments, reasonably scaled, are not *a priori* impossible. As it turns out the chemical transformations are in fact neither easier nor trickier to perform than in more traditionally set up experiments.

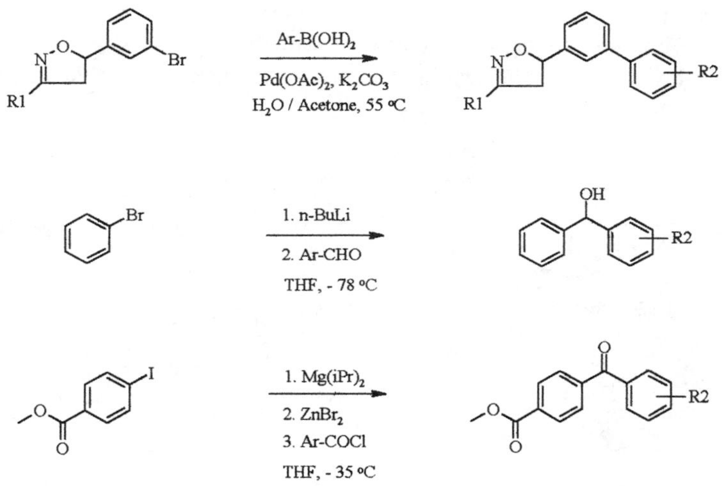

Figure 4: Chemistry Examples (performed)

Purification Concept

Problems in parallel synthesis obviously arise not from the fact of the synthesis itself but rather from the wide range of physical and/or chemical behavior of all the involved reagents and reactions. Since this is an intrinsic characteristic of any combinatorial approach, one is confronted with a typical management problem: namely to find the line between targeting maximum diversity and increased speed resulting from parallel handling. In order to obtain a maximum number of clean products, the chemist has either the possibility to study all the reaction parameter carefully and subsequently choose a reduced set of no more maximally diverse reagents (!) or give up the parallel approach altogether and dedicate a maximum of commitment and attention to the synthesis of any individual compound. A partial way out of this dilemma, namely the application of chromatographic methods, is of course not new but has more recently become a very attractive option because significant improvements in the area of automation and software control led to higher throughput. Hyphenation of liquid chromatography with mass spectrometry (LC/MS) allows the simultaneous separation and identification of compounds and is undoubtedly the technical solution most often used in industry to fulfill the objective of increased throughput (12,13). Such purification platforms are by definition not cheap, but an analysis of how chemists time is traditionally spent in most organic synthetic laboratories (Figure 5) shows the huge potential for a fast return on the capital investment without any sacrifice on behalf of their main task, which is the search for correlations between chemical structure and biological activity.

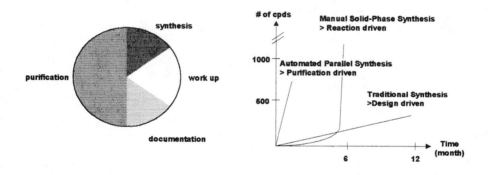

Figure 5: Saving Potential

We currently operate successfully several LC/MS systems in our research chemistry units across the whole research organization, i.e., at our research sites in Basel, Switzerland, as well as at RTP in the U.S. These purification platforms can be operated either in an analytical or preparative mode (Figure 6) and have overall significantly shortened the time to isolate and characterize compounds submitted for screening. Although it has to be clearly said that some limitations still exist, it is also true that these machines quickly become rather indispensable tools.

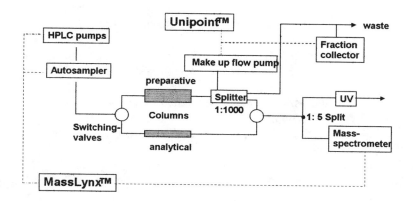

Figure 6: Wiring Diagram LC/MS

In summary, the *high speed synthesis* platform as just described, together with a sizable and high quality chemicals stockroom, an efficient electronic building block management, automated data analysis tools as described earlier (*14*) and a good mix of experienced and dedicated chemists are the basis of a very efficient hit-follow up process at Novartis. And most importantly, once the platform is in place and the various workstations are up and running, the chemists can again focus on what they like most: designing and preparing compounds in order to improve their biological activity.

References

1. Brown D. *Molecular Diversity*, **1996**, *2*, 217-222.
2. Furka, A.; Sebestyen, F.; Asgedom, M.; Dibo, G. *Int. J. Pept. Prot. Res.* **1991**, 37, 487-493.
3. Lam, K.S.; Salmon, S.E.; Hersh, E.M.; Hruby, V.J.; Kazmiersky, W.M.; Knapp, R.J. *Nature*, **1991**, *354*, 82-84.
4. Needles, M.C; Jones, D.G.; Tate, E.H.; Heinkel, G.L.; Kochersberger, L.M.; Dower, W.J.; Barrett, R.W.; Gallop, M.A. *Proc. Natl. Acad. Sci. USA* **1993**, *90*, 10700-10704.

5. Moran, E.J.; Sarshar, S.; Cargill, J.F.; Shabaz, M.M.; Lio, A.; Mjalli, A.M.; Armstrong, R.W. *J. Am. Chem. Soc.* **1995**, *117*, 10787-10788.
6. Bailey, N.; Cooper, A.W.; Deal, M.J.; Dean, A.W.; Gore, A.L.; Hawes, M.C.; Judd, D.B.; Merritt, A.T.; Storer, R.; Travers, S.; Watson, S.P. *Chimia* **1997**, *51*, 832-837.
7. Han, H.; Wolfe, M.M.; Brenner, S.; Janda, K.D. *Proc. Natl. Acad. Sci. USA* **1995**, *92*, 6419-6423.
8. Kaldor, S.W.; Fritz, J.E.; Tang, J; McKinney, E.R. *Bioorg. Med. Chem. Lett.* **1996**, *6*, 3041-3044.
9. Brown, R.D.; Martin, Y.C. *J. Chem. Inf. Comput. Sci.* **1996**, *36*, 572-584.
10. Lipinski, C.A.; Lombardo, F.; Dominy, B.W.; Feeney, P.J. *Advanced Drug Delivery Reviews* **1997**, *23*, 3-25.
11. Meyers, H.V.; Dilley, G.J.; Durgin, T.L.; Powers, T.S.; Winssinger, N.A.; Zhu, H.; Pavia, P.R. *Molecular Diversity* **1995**, *1*, 13-20.
12. Weller, H.N.; Young, M.G.; Michalczyk, S.J.; Reitnauer, G.H.; Cooley, R.S.; Rahn, P.C.; Loyd, D.J.; Fiore, D.; Fischman, S.J. *Molecular Diversity* **1997**, *3*, 61-70.
13. Zeng, L.; Kassel, D.B. *Anal. Chem.* **1988**, *70*, 4380-4388.
14. Diggelmann, M.; Gassmann, E.; Stämpfli, A. *Pestic. Sci.* **1999**, *55*, 374-376.

Chapter 18

The Application of Combinatorial Chemistry in Agrochemical Discovery

William A. Kleschick[1], L. Navelle Davis[1], Michael R. Dick[1], Joseph R. Garlich[1], Eric J. Martin[2], Nailah Orr[1], Simon C. Ng[2], Daniel J. Pernich[1], Steven H. Unger[2], Gerald B. Watson[1], and Ronald N. Zuckermann[2]

[1]Dow AgroSciences, 9330 Zionsville Road, Indianapolis, IN 46268
[2]Chiron Corporation, 4560 Horton Avenue, Emeryville, CA 94608

The challenge before the agrochemical discovery community is to take the learning from the pharmaceutical industry experience in combinatorial chemistry and adapt it to drive greater efficiency in our industry. This report outlines two examples of application of this technology in agrochemical discovery. The first example illustrates the screening of mixtures of N-substituted glycines to identify an insecticide lead. The second example involves the development of a library of individual compounds based on a 2-amino-4-aryloxypyrimidine template to optimize an insecticide lead.

Combinatorial chemistry is a proven and valuable tool for lead generation and lead optimization in the discovery of new pharmaceutical agents. The number of clinical candidates that owe their existence, at least in part, to combinatorial chemistry technology has grown in recent years (*1*). By comparison the use of the technology in an agrochemical discovery context is in its infancy. The challenge for the agrochemical discovery community is to adapt the technology to meet the needs of our environment. The agrochemical discovery environment is different from that of the pharmaceutical industry in that it relies heavily on whole organism screening. Understanding where the technology offers significant advantage in addressing our need to become more efficient and productive is critical to our success.

We have been exploring opportunities to use the technology in both lead generation and lead optimization. We have used both collaborations with companies with a strong technology base in combinatorial chemistry and in-house efforts to accelerate our learning process. In this report we document two examples of using the technology to our advantage. The first example illustrates the screening of mixtures of N-substituted glycines to identify an insecticide lead. This project was pursued in collaboration with Chiron Corporation. The second example involves the development of a library of individual compounds based on a 2-amino-4-aryloxypyrimidine template to optimize an insecticide lead.

N-Substituted Glycines

In 1992 scientists at Chiron Corporation reported the development of oligomers of N-substituted glycines as analogs of peptides (Figure 1) (2). The N-substituted glycine or peptoid analogs of several known peptides were shown to possess comparable binding affinities. The peptoids were prepared by an innovative submonomer solid-phase synthesis (3). The approach involves successive acylations of an amine functionality on a polymer resin with bromoacetic acid followed displacement of the bromine with an amine (Figure 2).

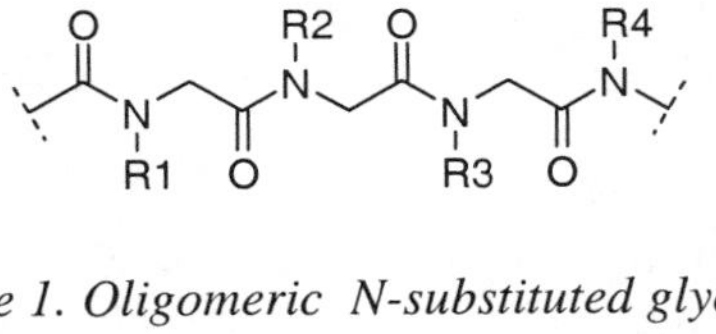

Figure 1. Oligomeric N-substituted glycines.

Through a collaborative agreement with Chiron, we screened approximately 350,000 peptoids in a number of assays including a competitive binding assay using $[^3H]$-α-bungarotoxin against the insect nicotinic acetylcholine receptor. The compounds were prepared by a split and pool method and were obtained as mixtures (a total of 151 pools) containing as few as 24 dimers per pool up to

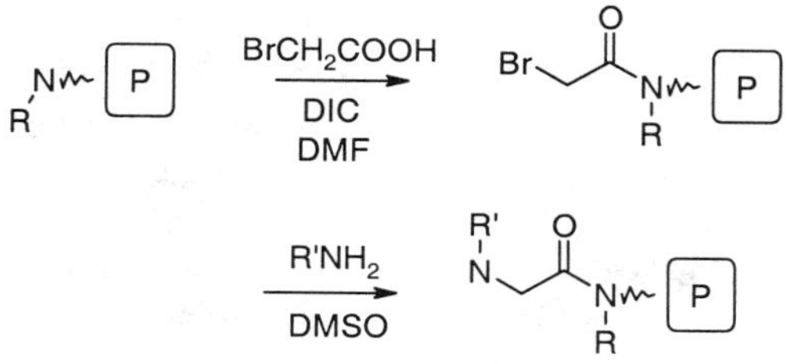

Figure 2. Sub-monomer synthesis of N-substituted glycine units.

13,824 tetramers per pool. In the initial screen five pools exhibited significant displacement of [^{3}H]-α-bungarotoxin at a concentration of 1 μM or below. One pool exhibited the greatest level of displacement in the initial assay pool and a good dose response curve in follow-up assays. This pool (11762) containing 576 trimers (24 R2 groups x 24 R3 groups) was chosen for deconvolution.

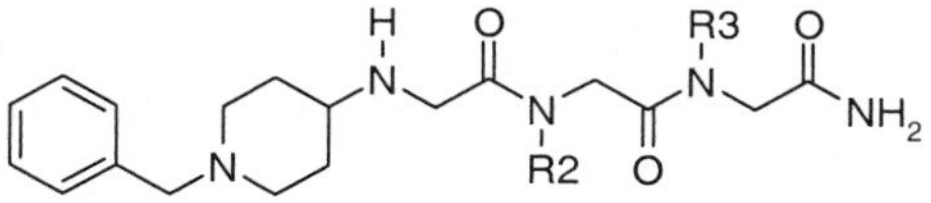

11762

Deconvolution of pool 11762 resulted in 24 pools containing 24 peptoids each with the same R2 group and all 24 R3 groups. Two of the sub-pools (21381 and 21384) exhibited > 50% displacement of [^{3}H]-α-bungarotoxin and a good dose response. Sub-pool 21381 had an apparent K_i of 1.5 μM in displacement of [^{3}H]-α-bungarotoxin, but did not show significant displacement of [^{3}H]-imidacloprid. Sub-pool 21384 had an apparent K_i of 0.6 μM in displacement of [^{3}H]-α-bungarotoxin, and also displaced [^{3}H]-imidacloprid with an apparent K_i of 2 μM. Sub-pool 21384 was chosen for final deconvolution based on these results.

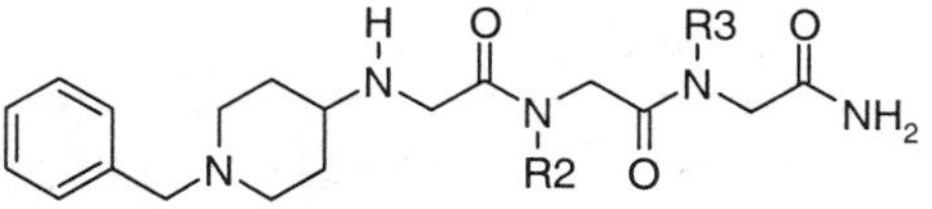

21381: R2 = Me
21384: R2 = H

Deconvolution of sub-pool 21384 provided 24 individual compounds. One of these compounds (27054) was found to cause 50% displacement of [^{3}H]-α-bungarotoxin at 0.27 μM. Compound 27054 was also found to be a nicotine antagonist (IC$_{50}$ = 0.98 μM) and possessed weak insecticidal activity on *Heliothis virescens* by injection. No activity was observed in dietary assays. Unfortunately, analog synthesis around 27054 using conventional and combinatorial methods did not yield a significant improvement in activity (*4*).

27054

2-Amino-4-aryloxypyrimidines

During the course of screening compounds for insecticidal activity, 4-(2,4-dichlorophenoxy)-2-isopropylamino-6-trifluoromethylpyrimidine surfaced as a lead compound with good activity against rice pests. This lead seemed ideally suited for optimization using combinatorial chemistry methods *via* a route represented in a retrosynthetic analysis depicted in Figure 3. The amino diversity element was envisioned to be introduced by nucleophilic aromatic substitution of a sulfonyl group attached to a polymer resin. The requisite substrate could be derived from pyrimidine ring synthesis on the resin starting from the appropriate thiourea functionalized resin or by attachment of a preformed 2-mercaptopyrimidine to a resin. In practice, building the pyrimidine ring on the resin proved to be difficult. The ultimate synthetic protocol was accomplished starting from the appropriate 2-mercaptopyrimidine and is outlined in Figure 4. Workers at Signal Pharmaceuticals have reported a similar solid-phase approach to introducing an amino diversity element at the 2-position of a similar pyrimidine substrate (*5*). In addition, researchers at Hoffman-La Roche have reported a successful synthesis of a related pyrimidine substrate on a solid support (*6*). Their approach involves a cyclocondensation of an acetylenic ketone with a resin-bound thiouronium salt.

The protocol in Figure 4 was used to prepare a library of 96 compounds based on 8 amine and 12 phenolic diversity elements (Figures 5 and 6).

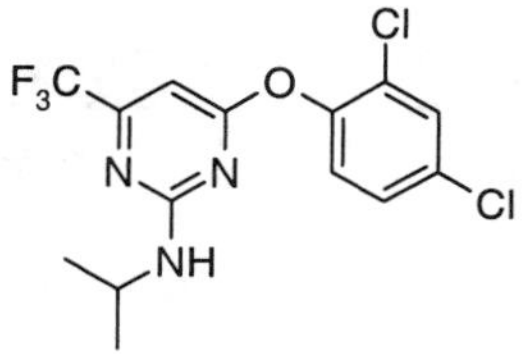

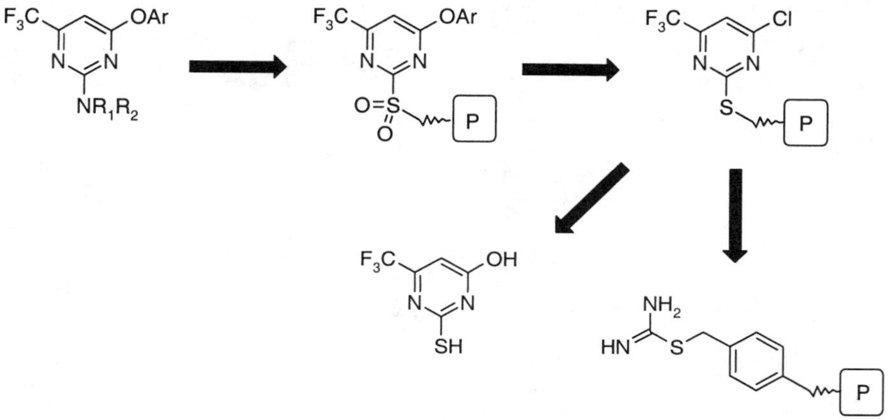

Figure 3. Retrosynthetic approaches to 2-amino-4-aryloxypyrimidines.

4-hydroxy-2-mercapto-6-trifluoromethylpyrimidine was attached to Merrifield resin through alkylation of sulfur. The hydroxy group of the resin-bound substrate was converted to chlorine by treatment with POCl₃, and the chlorine was displaced with the sodium salt of the phenolic diversity element. This reaction was run in a mixture of 1,2-dimethoxyethane (DME), tetrahydrofuran (THF) and N,N-dimethylformamide (DMF) to ensure solubility of the sodium salt of the phenol. The sulfur was oxidized with *m*-chloroperoxybenzoic acid (*m*-CPBA). The resulting sulfoxide was cleanly displaced with the appropriate primary or secondary amine to introduce the second diversity element and effect cleavage from the resin. HPLC and GC/MS were used to determine the purity of each of the crude reaction products. Seventy-five percent (75%) of the samples had a purity equal to or greater than 50%, 68% had a purity equal to or greater than 75%, and 41% of the samples had a purity equal to or greater than 90% (Table 1). Low yields were associated with the sterically hindered diisopropylamine (**H**) and the phenolic inputs N-(3-hydroxyphenyl)pyrrolidine (**8**), 7-hydroxycoumarin (**10**) and 5-chloro-8-hydroxyquinoline (**11**). In seven instances none of the desired product was detected in the crude reaction product.

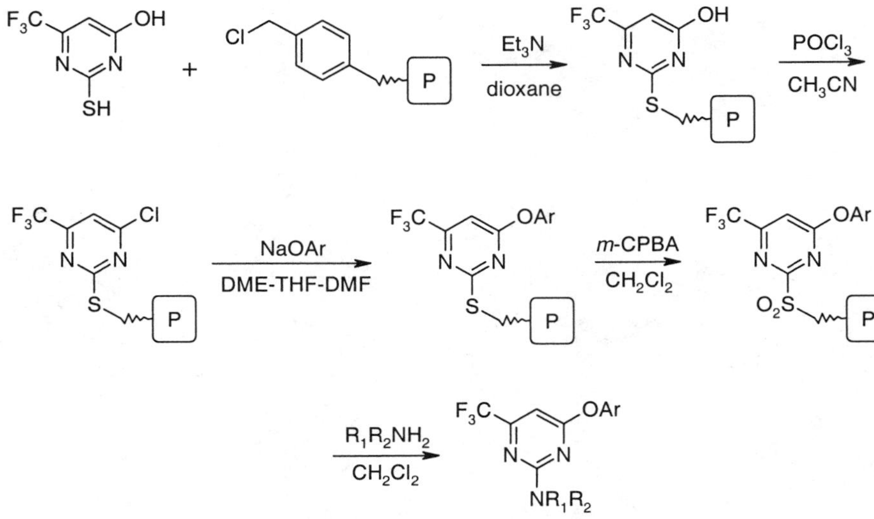

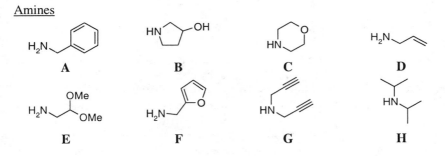

Figure 4. Solid-phase synthesis of 2-amino-4-aryloxypyrimidines.

Amines

A

B

C

D

E

F

G

H

Figure 5. Amine diversity elements

The 96 compound library was screened against four insect pests (*Heliothis virescens*, *Spodoptera exigua*, *Meloidogyne* spp. and *Lipathis erysimi*) in dietary assays. Six of the compounds showed detectable levels of activity against at least one of these insect pests (**C3**, **F4**, **A9**, **G9**, **G12**, and **H12**). However, attempts to further optimize the activity through the synthesis of additional libraries were unsuccessful.

Phenols

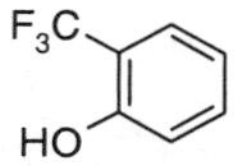

Figure 6. Phenolic diversity elements.

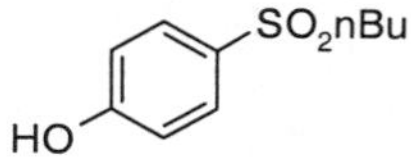

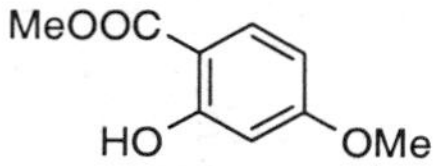

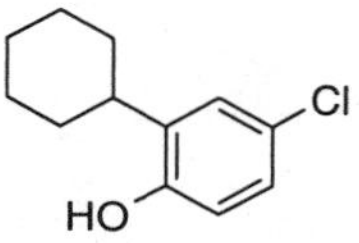

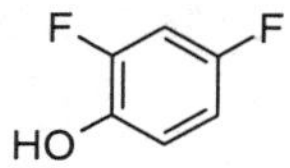

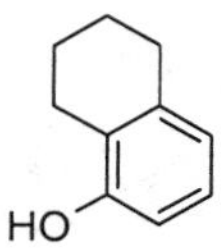

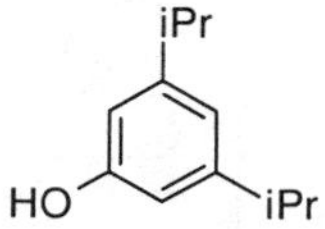

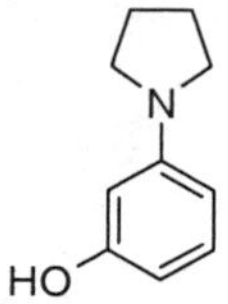

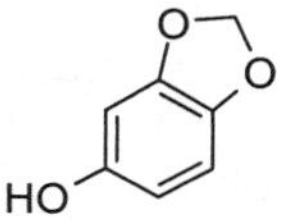

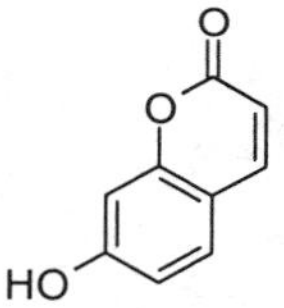

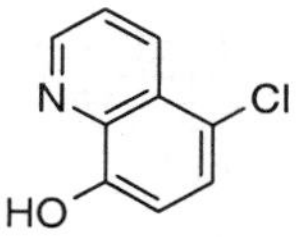

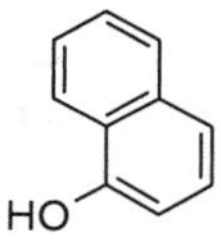

    In the absence of a significant breakthrough in activity, we have looked for ways to quantify the impact of our efforts in combinatorial chemistry. One approach we pursued was to assess the value of the information content in the screening results of libraries such as the 2-amino-4-aryloxypyrimidine library. The time required to develop and execute the solid-phase protocol for the synthesis of the 96 compound library was approximately 3 months. We estimate that an average chemist would have synthesized no more than 20 compounds using conventional synthetic methods in the same time frame. We then assessed the probabilities of finding 1, 2, 3, 4, 5 or all 6 of the actives found in a random set of 20 compounds within the 96 compound library. The probability of finding 1 or 2 of the actives is relatively high (i.e., approximately 40% and 25% respectively). However, the chances of finding 3, 4, 5 or all 6 of the actives were considerably lower (i.e., approximately 1 in 12, 1 in 67, 1 in 769 and 1 in 25,000 respectively). The value of the information present in all six data points to reach conclusions or to plan follow-up experiments is considerably higher that that present in a few. As a consequence the resource investment to prepare this

Table I. Purity of Compounds in a 2-Amino-4-aryloxyphenoxypyrimidine Library

	A	B	C	D	E	F	G	H
1	92%	98%	93%	96%	99%	92%	92%	35%
2	94%	85%	90%	95%	96%	95%	75%	50%
3	96%	96%	95%	96%	81%	96%	84%	50%
4	79%	78%	75%	86%	76%	80%	85%	40%
5	98%	97%	90%	96%	82%	95%	83%	25%
6	95%	96%	92%	92%	91%	90%	92%	64%
7	82%	90%	75%	83%	77%	88%	85%	10%
8	20%	0%	14%	80%	0%	48%	18%	29%
9	80%	85%	83%	90%	85%	88%	12%	61%
10	90%	0%	7%	0%	0%	9%	0%	0%
11	74%	60%	23%	85%	76%	71%	23%	30%
12	95%	90%	92%	95%	97%	90%	92%	30%

96 compound library is more than justified. This advantage is likely to be magnified in instances where resource investments in protocol design are leveraged to produce larger libraries.

Conclusion

We have demonstrated that combinatorial chemistry methods can be effective in generating new leads and of value in optimizing leads found by other means. The example of screening N-substituted glycine libraries to identify an insecticide lead highlights the value of screening pools of compounds in an *in vitro* assay. While working with pools of compounds has become less popular in recent years, the combination of pool and split methods with a highly discriminating *in vitro* assay can be highly efficient. The example of the 2-amino-4-aryloxypyrimidine library illustrates the value of the information content in a large collection of analogs compared to a more conventional series of analogs. As the experience base in these methods grows in the agrochemical industry, we expect that a track record of success similar to that of the pharmaceutical industry will be established.

References

1. Brown, R. K. *Modern Drug Discovery* July/August 1999, p 63.
2. Simon, R. J.; Kania, R. S.; Zuckerman, R. N.; Huebner, V. D.; Jewell, D. A.; Banville, S.; Ng, S.; Wang, L.; Rosenberg, S.; Marlowe, C. K.; Spellmeyer, D. C.; Tan, R.; Frankel, A. D.; Santi, D. V.; Cohen, F. E.; Bartlett, P. A. *Proc. Natl. Acad. Sci.* **1992**, *89*, 9367.
3. Zuckermann, R.N.; Kerr, J.M.; Kent, S.B.H.; Moos, W.H. *J.Am.Chem.Soc.* **1992**, *114*, 10646.
4. Garlich, J. R..; Orr, N.; Watson, G. B., unpublished results.
5. Gayo, L. M.; Suto, M. J. *Tetrahedron Lett.* **1997**, 211.
6. Obrecht, D.; Abrecht, C.; Grieder, A.; Villalgordo, J. M. *Helv. Chem. Acta* **1997**, *80*, 65.

Chapter 19

Combinatorial Chemistry in the Agrochemical Discovery Process at Zeneca

Richard D. Gless[1], Karl J. Fisher[1], and Donald R. James[1]

Zeneca Agricultural Chemicals, Western Research Center,
1200 So. 47th Street, Richmond, CA 94804
[1]Current address: Cambridge Discovery Chemistry,
1391 So. 49th Street, Richmond, CA 94804

Although combinatorial chemical methods have been employed in the pharmaceutical industry for a number of years, use of these techniques has only recently gained favor in agrochemical companies. Recent developments in the methods used to generate test compounds for Zeneca's agrochemical screen at both the US and UK research centers are summarized. Library design philosophy, philosophy and types of automation, as well as examples of chemistries carried out at Zeneca's US facility are discussed.

Introduction

A combinatorial chemistry discovery program has two prerequisites: first, the ability to produce large numbers of compounds, which usually implies some type of automated synthesis, and second, and more importantly, a high throughput screen which allows the evaluation of the compounds prepared. Zeneca has been among the first agrochemical companies to develop high throughput screens requiring small amounts of test substance, and, as a result, has had a vested interest in developing automated synthesis programs (*1-3*).

The history of automated synthesis at Zeneca dates to 1990-1991, when Zeneca Pharmaceuticals began a collaboration with Zymark to develop a robotic system for automated chemical synthesis. The first robot designed specifically for chemical synthesis was installed at the Zeneca Pharmaceuticals site in Alderley Park, UK, in late 1991, with the express purpose of improving overall chemist efficiency by allowing multiparallel synthesis of 50-100 test materials in *ca.* 100 mg quantities. Based on the success of this system, additional robots were installed, and, in early 1995, a similar robotic system was brought into the Zeneca Agrochemical Research Center at Jealott's Hill, UK. Manual combinatorial chemistry using a mixture strategy was initiated at the Western Research Center in Richmond in 1995, and in early 1997, a robotic system similar to that in the UK was placed at the US site. Zeneca established a combinatorial chemistry group in the UK in mid 1996 to coincide with the initial operation of one of the first high throughput agrochemical screens. At the end of 1997, a second combinatorial chemistry group was placed at the US research center. The high throughput screen became fully operational in late 1998.

The use of combinatorial chemistry for lead generation differs in many ways from the traditional approach to agrochemical discovery. Traditionally, the discovery chemist would explore an idea by preparing up to about twenty-five compounds, then wait for screening results. If activity was found or the idea was still intriguing, the chemist would make a few more compounds or move on to another idea. Automated multiparallel synthesis allows the chemist to increase the number of compounds prepared by a factor of 10, but high throughput screening demands that the chemist consider the preparation of thousands, tens of thousands, or even larger numbers of compounds which necessitates the use of combinatorial chemical techniques. The preparation of combinatorial libraries has multiple components that must be considered including template and library design, chemical logistics and automation, analysis and quality control, and data management. This paper will discuss the first three of these areas.

Library Design Considerations

Probably the most important aspect of the combinatorial chemistry process is library design (*4-6*). Making lots of compounds and getting lucky was considered a valid strategy in the early days of combinatorial chemistry and, with some large libraries, may still be a valid strategy. However, with increasing sophistication and the availability of more powerful computational methods, both large and small libraries are now designed to be "diverse" at least based on various computational criteria. At present the emphasis in the pharmaceutical area appears to be away from large libraries to "focused" libraries that are targeted towards specific physical properties and/or biological targets.

While libraries can be designed in many different ways depending on the target, there are several considerations that are or should be common to most if not all design schemes. Libraries typically have a core template or scaffold, either designed or with known biological activity. Libraries may be either "product-like" or "lead-like". In the former case subsequent modification of actives involves substitution of one diversity element for another to modify physical or biological properties. In the latter case substituents or functionality is added to accomplish the same result. Diversity and novelty are prime considerations. These are normally determined by evaluating overlap with the current company collection, known chemical entities, patented materials, and commercial compound sources, which might be avoided to a varying extent depending on the purpose of the library. Physical properties (e. g., calculated log P, molecular weight, template and substituent H-bond acceptor/donor functionality, etc.) are usually considered. In many cases it is felt that at least three significant binding groups should be present. Library products, especially in lead finding libraries should possess appropriate conformational flexibility to allow maximum potential for binding to target sites. The logistics of library production must also be considered. A robust, efficient chemical route compatible with multiple, diverse substituents must be available, as well as a collection of appropriate diversity reagents.

At Zeneca three variations on template design are employed. The first approach, which is very common in combinatorial chemistry, is the use of a "core structure". This "core structure" might be agchem-like, protein-like, or a "biorational" substructure. It might be natural product based or physical property based, or it might be based on interesting or "easy" chemistry. Or, and this has been a common approach, the library scaffold might be based on a novel heterocycle or a variation on a known or uncommon heterocycle. A second approach is to attach two or more agchem-like or other substructures, containing binding groups with appropriate functionality, with various "linkers" usually consisting of two to seven atoms. A third approach is to explore binding in geometric space, e.g., a "head group" or "warhead" is attached to an aromatic ring or other core structure which can be substituted with appropriate functionality, in the case of an aromatic ring, in the ortho, meta, and para positions.

For libraries targeted at novel screen leads, library members should be selected to provide an optimum coverage of diversity space. In other words the compounds prepared should be selected to be as different from each other as possible to keep from making large numbers of what are essentially the same compound. Traditionally, when a discovery chemist set out to prepare test compounds around an idea, he would make a few analogs, perhaps a few substituted aromatic analogs and a few aliphatic analogs, as well as perhaps one or more simple heteroaromatic analogs. The typical diversity space coverage might look something like that shown in Fig. 1 with the "vein" of activity as shown.

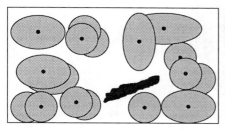 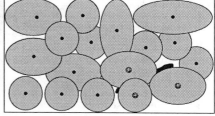

Fig. 1: Poor Diversity Space Coverage   *Fig. 2: Good Diversity Space Coverage*

A library that is well designed from the standpoint of diversity space coverage might look like the one shown in Fig. 2. Based on a diversity argument, one would expect more success using this second design. Diversity profiling is currently done at Zeneca using a Tanimoto coefficient comparison for both internal and external diversity, that is, a measure of how similar each member of the library is to others in the library and how similar each member is to compounds in the current corporate collection as well as compounds in a database of commercial and experimental pesticides. Having compounds whose similarity is less than 0.85 is considered a minimum desirable level for lead generation libraries.

Insuring that a library has appropriate physical properties is usually accomplished by comparison with some standard collection of compounds. Comparisons of molecular weight and calculated log P of members of a library with an in-house database of commercial and experimental pesticides is commonly used. Additional profiling methods for hydrogen donor and acceptor sites, rotational bonds, *etc.*, are under development at Zeneca. Profiling is especially useful for libraries directed at specific targets. An example of a calculated log P profile for one lead generation library *vs.* that of a in-house database of commercial and experimental pesticides, is shown in Fig. 3. Reasonable overlap with the calculated log P distribution for a selected set of commercial and experimental pesticides is present.

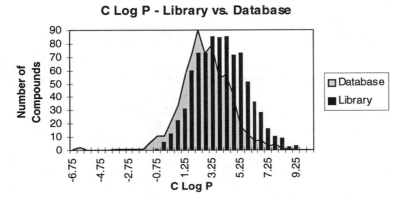

*Fig. 3: Comparison of calculated log P profile for one lead generation library
vs. In-house Pesticide Database*

Traditionally, combinatorial chemistry libraries have consisted of
compounds bearing/containing all possible combinations of substituents. Thus if
a template had three points of diversity, and there were ten reagents available for
each point of diversity, a 10 x 10 x 10 library of 1000 compounds was generated.
This approach is referred to as a "saturated matrix" library within Zeneca. It is
important to note that this does not mean that "diversity space" is saturated, but
only the matrix of available reagent combinations. Libraries prepared on solid
phase at the UK research center have, to date, used this approach. An alternate
approach employed at the Western Research Center is to make "sparse matrix"
libraries which are subsets of "saturated matrix" libraries. A "sparse matrix" can
be either a "regular sparse matrix" or a "computed sparse matrix". In a library
containing two diversity substituents, typical of Zeneca solution phase libraries
to date, a "regular sparse matrix" is generated from 10 different substituents A
and 10 different substituents B by pairing every third B with each A to given a
33 compound library from a virtual library of 100 possible products. A
"computed sparse matrix" library is generated in the same way except that
substituents A and B are selected by Tanimoto comparisons to afford products
which are dissimilar at some specified level. Fig. 4 shows the three library
design types assuming 10 A substituents (columns) and 10 B substituents (rows).
Since the Tanimoto coefficient primarily measures atomic differences,
comparisons based on this metric tend to overemphasize elemental identity over
variations in substitution patterns or geometry. As a result a "calculated sparse
matrix" chosen using the Tanimoto coefficient as the selection metric may pick a
relatively large number of examples bearing, *e. g.*, a trifluoromethyl group or a

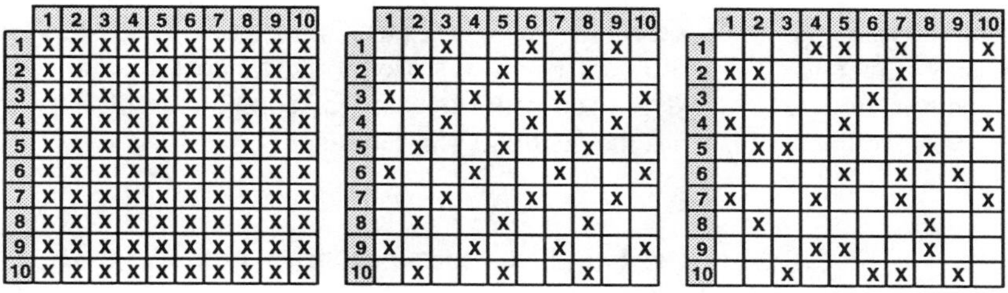

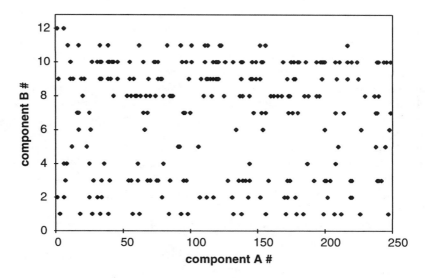

Fig. 4: "Saturated Matrix", "Regular Sparse Matrix", and "Computed Sparse Matrix" Substituent Pairings

small heterocycle, at the expense of simpler structures. Fig. 5 shows the pairings of substituents A and B selected using the Tanimoto coefficient as the selection metric in an example library. B components 3, 8, 9, and 10 are overemphasized relative to the rest of the library. To compensate for the tendency of the "calculated sparse matrix" to overselect certain reagents, libraries at Zeneca's Western Research Center are typically composed of an equal number of

Fig. 5: Pairings of substituents A and B selected using the Tanimoto coefficient as a selection metric in an example library

members selected as a "calculated sparse matrix" and as a "regular sparse matrix." Since the Tanimoto coefficient was used to help select diverse sets of reagents in both the A and B lists, using it to select compounds in the entire library as well only magnifies the method's shortcomings.

Chemical Logistics and Automation

There are several practical considerations involved in the preparation of combinatorial chemistry libraries. Obviously, one must start with chemistry that is attractive from the standpoint of the library design considerations stated above, but there are also the additional logistics of selecting, tracking, and validating library reagents. The selection process usually involves some screening of the Available Chemicals Directory™ or other databases, picking reagents based on appropriate composition, molecular weight, cost, toxicity, *etc.*, performing some sort of clustering analysis, *etc.*, and repeating this process iteratively until a manageable set of reagents is defined and purchased. The chemical robustness of the synthetic route must be confirmed, and some reagents may be eliminated for lack of reactivity, or the library chemistry may be modified to accommodate the range of reagents selected. A virtual library must be generated and structures enumerated. Intermediates must be prepared in sufficient quantity and purity, a step that can be a major bottleneck. Reagents must be validated to confirm that they have the correct structure and react under the conditions being employed. And finally the nontrivial realities of automated production and quality control must be addressed.

Automation of chemical synthesis is usually viewed as three primary operations, each of which can be further divided into sub processes: a) running reactions (taring tubes, weighing starting materials, dispensing reagents, heating and agitating reactions), b) working-up reactions (evaporating reaction solvent, dispensing work-up solvents, partition, evaporation, weighing products), and c) preparing samples (dispensing solvent, dissolution, aliquoting samples). All of these processes are conveniently carried out on the standard Zymark synthesis robot used within Zeneca. This robot includes the basic Zymark platform (Fig. 6) and a Savant SpeedVac for performing solvent evaporations.

The Zymark synthesis platform as modified for the needs of Zeneca's Western Research Center (7) consists of a Zymark XP robot with a GP hand as well as appropriately sized syringe hands, a Mettler balance accurate to *ca.* 0.3 mg, three Master Laboratory Stations (MLS units) for addition of up to eight reagents/solvents as well as cannula rinse solvent and an inert atmosphere sparge, a capping/uncapping station, a reagent dispensing station, a cannula station for effecting phase separations, a Stem heater/stirrer, and various racks for holding a variety of reaction test tubes, product and analytical sample vials, and 96 well plates.

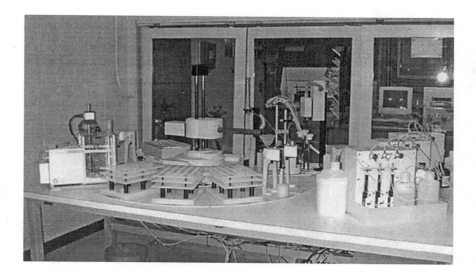

Fig. 6: Zymark Synthesis Robot

The Zymark system has proven to be very successful at accomplishing its stated objectives, namely, providing rapid analog synthesis, reproducing the majority of synthetic chemistry undertaken in the laboratory with minimum modification, and increasing chemists' synthetic efficiency by allowing synthesis of 50-100 compounds in parallel in *ca.*100 mg quantities. During the first year of operation (1997) at the Western Research Center, standard robotic procedures and QC methods (automated chemical ionization MS) were developed that allowed the preparation of 100 compounds/week with *ca.* 2,800 samples being prepared, at first in 4 dram vials and later in 96 well plates. This was a commendable accomplishment given that the first runs were carried out in late February and production was decidedly reduced during the last three months of the year as a result of a major site reorganization. Typical chemistries included amide couplings, Michael additions, C-acylations of 1,3-diketones, alkylations of O, S, N, and $RNHNH_2$, as well as preparations of esters, sulfonamides, carbamates, and pyrazoles.

In spite of this success, the Zymark system was basically not designed to handle the small volumes involved in the preparation of samples for the newly instituted high throughput screen. A second system based on the Gilson 215

Liquid Handler was assembled to provide a system that overcame these limitations (8). The system (Fig. 7) consists of the following modules: a Gilson 215 Liquid Handler, Savant SpeedVac, Glas-Col Multi-Pulse Vortex, Robbins Hydra96, custom hot plates based on a Cole Parmer nine position stirring hotplate, and the existing Zymark robot. All solution handling (reagent and solvent dispensing, solids dissolution, analytical sample and 96 well plate preparation) is carried out using the Gilson 215 Liquid Handler with a minimum

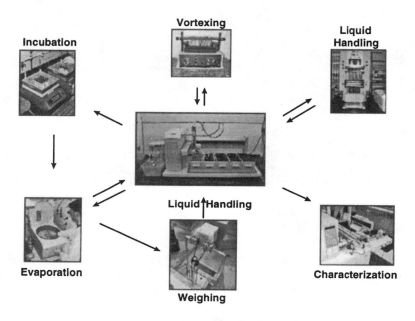

Fig. 7: Gilson 215 Based Synthesis System

number of deck configurations. Solvent evaporations are carried out on the Savant SpeedVac with a custom rotor holding up to ninety-six 16 mm x 100 mm test tubes or a standard rotor accommodating up to four 96-well plates. Reactions requiring heating above *ca.* 80°C are carried out in glass test tubes in custom aluminum heating blocks on the Cole Parmer hotplates using a small magnetic stirrer in each tube. Reactions in 96-well plates are agitated on a Labnet Shaker-30, or, where appropriate, using a Gilson 818 Automix. Test tube reaction products are worked-up using liquid-liquid extraction with agitation provided by a multiposition Glas-Col Multi-Pulse Vortex, while 96-well plate reaction products are purified with either SPE or treatment with scavenger resins. Tube weighing operations are carried out on the existing Zymark Robot. The Gilson 215 based system affords excellent small volume liquid handling and has the additional advantages of fitting in a standard fume hood and using a more modern programming interface. The modularity of the system allows easy modification and debottlenecking. A text based menuing system allows users to select from a list of standard operations.

The Gilson 215 based modular system is popular with the chemists and has proven to be quite productive. From 150 - 300 samples per week are easily prepared with a record of 1300 compounds in one week in 96 well plates by one chemist. Production has progressed from 75 reactions per run in tubes to reactions in 96 well plates. Both solution and solid phase methods (using Robbins Blocks) have been carried out. Reactions carried out to date include amide couplings, urea formation, peracid oxidations, $NaBH_4$ reductions, O, S displacements of aromatic halogen, and amidine formation.

Analysis and Quality Control

Quality control of product libraries is a major concern given the considerable amount of resource that can be expended on design and synthesis of libraries as well as on the initial follow-up of active library compounds. A number of quality control options are available including full characterization of all samples, partial characterization of all samples, or partial characterization of some statistically valid subset of the library. Full characterization of all samples is normally considered excessive for samples being prepared for high throughput screening given the large numbers of products and the physical task of analyzing and archiving the data, but may be appropriate for follow-up samples or smaller lead following libraries.

At Zeneca's Western Research Center all samples are analyzed in initial production runs by mass spectroscopy using desorption chemical ionization mass spectroscopy (CI-MS) on a Micromass Trio 2000 Quadrupole Mass Spectrometer. As confidence with the chemistry used in building a library grows, MS analysis is carried out on some subset of samples. Currently MS analyses are performed on either all samples or all odd samples with HPLC analysis and NMR being carried out on *ca.* 5% of each library. NMR methods using a Nalorac flow cell and solvent suppression are under development to replace manual NMR. At the UK research center, analysis of large solid phase libraries is carried out by characterization of a *ca.* 10% subset of the library by LC/MS.

Evaluation of mass spectral data at the US facility is carried out using a fully automated system developed in-house. Analytical input in the form of chemical formula data for each reagent and product is fed into the analysis system, and an analytical evaluation summary of all samples in each robot preparation is summarized in a one page printout. Additional details including a summary of the data used for the automated evaluation, *i. e.*, total ion current, ion current for the peak of interest as well as MH+ and MNH$_4$+, calculated and observed isotope ratios for peaks of interest, and the actual mass spectra trace are also prepared as a single page document.

Results

Although the majority of successes of Zeneca's combinatorial chemistry efforts are still propriety, no discussion of a chemical strategy is complete without some chemical structures. The products shown below are a few of the structures generated early in the library synthesis program that are of biological interest (9).

Phenoxypyrazole 1 exhibited 2,4-D-like symptomology primarily on broadleaf weeds with X = Cl, H. It was an additive hit with activity ranking Alk = CH$_2$COOEt > (CH$_2$)$_5$COOEt > CH(CH$_3$)COOEt > (CH$_2$)$_4$COOEt. Imidazole 2 showed post emergent broadleaf herbicide activity. Curiously, out of the sixty analogs prepared, only the cyclopropylmethyl product showed any activity. Six mixtures of ten analogs of thioether 3 were prepared; five were active, but only two were sufficiently active to warrant deconvolution. The most active herbicides were substituted benzyl ethers such as 4-fluoro or 3-trifluoromethyl. Methyl triazole 4 showed leaf burn symptomology on grasses when Alk = 3,4-

dichlorobenzyl. Phenyl triazole 5 proved to be a bleaching herbicide with Alk = 4-fluorobenzyl being one of the more active compounds. These bleaching herbicides do not inhibit phytoene desaturase. Triazolethioacetamide 6 was isolated as the single active component from a mixture of ten compounds. It shows both pre and post-emergent activity at 1 Kg/ha. This lead compound has a reasonably simple and novel structure. Biochemical studies indicate it probably represents a new mode of action.

References

1. Brown, D. *Molecular Diversity*, **1996**, *2*, 217-222.
2. Bailey, N; Cooper, A. W. J.; Deal, J.M.; Dean, A. W.; Gore, A. L.; Hawes, M. C.; Judd, D. B.; Merritt, A. T.; Storer, R.; Travers, S.; Watson, S. P. *Chimia,* **1997**, *51*, 832-837.
3. Balkenhohl, F.; von dem Bussche-Hunnefeld, C.; Lansky, A.; Zechel, C. *Angew. Chem. Int. Ed. Engl.*, **1996**, *33*, 2288-2337.
4. van Drie, J.H.; Lajiness, M. S. *Drug Discovery Today* **1998**, *3*, 274-283
5. Martin, E. J. *J. Comb. Chem.* **1999**, *1*, 32-45.
6. Ghose, A. K.; Viswanadhan, V. N.; Wendoloski, J. J. *J. Comb.Chem.* **1999**, *1*, 55-68.
7. Gless, R. D. "Evolution of Automated Synthesis Methodology in Zeneca's Agchemical Discovery Process" presented at the International Symposium on Laboratory Automation and Robotics in Boston, MA, Oct. 17-20, 1999.
8. Gless, R. D.; James D. R.; Cornell, J. L.; Neal, J. R.; Keller, M. T.; McKenna, M. M.; Partridge, L. G.; Fitzgerald, B. A.; Bowler, D. J. "Solution Phase Synthesis of Combinatorial Libraries using a Gilson 215 Based Modular Robotic System in Zeneca's Agchemical Discovery Process", presented as a poster at the International Symposium on Laboratory Automation and Robotics in Boston, MA, Oct. 17-20, 1999.
9. Fisher, K. J.; Felix, R. A.; Oliver, R. M. "Screening Mixtures: An Experiment in Pesticide Lead Generation., presented at the 217th National Meeting of the American Chemical Society in Anaheim, CA, Mar. 21-26, 1999.

Chapter 20

Biorational Design, Synthesis, and Inhibition of Photosystem II Inhibitors

Huayin Liu, Yinlin Sha, Aiming Yu, Huifen Tan, and Huazheng Yang[1]

State Key Laboratory of Elemento-Organic Chemistry, Institute of
Elemento-Organic Chemistry, Nankai University, Tianjin 300071, China
[1]Corresponding author.

Molecular modeling of cyanoacrylates(cyanoacrylamides) with
D1 protein of *Pisum sativum* have been presented. Studies
show that the binding force includes mainly: H-bond
interaction, Van der Waals and π-ring stacking interaction. It
was found that SER 268 in D1 protein might be an important
binding site. Thus some new cyanoacrylates (cyanoacrylamides) were designed and synthesized. For rapid optimization,
combinatorial methods were introduced to synthesize 3-N-
substituted (2-thio)hydrouracils library by the acidic cyclization-cleavage of ureas and thioureas from Wang resin. Their
Hill inhibition is discussed.

Photosynthesis is a special and important physiological and biochemical
phenomenon for plants. Hence the selection of photosynthesis as the herbicidal
target is desirable to get nontoxic pesticides. Recently, Deisenhofer has
successfully elucidated the three dimensional structure of L protein of the
photosynthetic reaction center of *Rps. Viridis* with X-ray diffraction(1), and this
important result makes it possible to design new photosystem II (PS II) inhibitors
based on the receptor structure.

A large number of inhibitors are known to block electron transport in PS II
by displacing plastoquinone from Q_B-binding niche of the D1 protein in the

reaction center(1,2). Based on the three dimensional structure of L protein of photosynthetic reaction center of *Rps. Viridis*, the 3D-structure of *Pisum sativum* 32Kdal(D1) was constructed by homology modeling method, and the properties, such as steric, hydrophobic, electrostatic properties of active sites region of the D1 protein were also characterized by the grid search method with probes of methyl group, H_2O, amino and proton. For molecular modeling , the software SYBYL(Version 6.2) from Tripos Associates Inc. was used(3,4).

Cyanoacrylates are potent inhibitors of photosynthetic electron transport. A number of studies concerning the inhibition of photosynthetic electron flow in PS II with a series of cyanoacrylate inhibitors have shown that the potency of cyanoacrylates in blocking photosynthetic electron flow is extremely sensitive to minor structural variation(5,6). The structures and inhibition(pI_{50} value) of cyanoacrylates(cyanoacrylamides) are shown in Table 1. It was considered that these inhibitors would be useful probes for determining the nature of the receptor topography.

Table 1 The structures and inhibition(pI_{50} value) of cyanoacrylates(6)

$$R^1NH\diagdown C=C\diagup COXR$$
$$Et \qquad CN \qquad \mathbf{1}$$

No.	R^1	XR	pI_{50}
1a	p-ClC$_6$H$_4$	OC$_2$H$_4$OEt	5.70
1b	p-ClC$_6$H$_4$CH$_2$	OC$_2$H$_4$OEt	8.20
1c	p-ClC$_6$H$_4$CH$_2$	NHC$_2$H$_4$OEt	

Thus, based on the constructed receptor model, cyanoacrylates(cyano-acrylamides) were selected for molecular modeling in order to find additional active sites in D1 protein.

Binding of Cyanoacrylates(Cyanoacrylamides) with D1 Protein

Plate 1 shows compound **1b** in the Q_B binding pocket of D1 protein. It appears that the oxygen atom of the ester carbonyl group in cyanoacrylates binds to the hydroxyl of SER 264 residues in agreement with the literature(7,8), and the nitrogen atom in cyano group binds to LEU 271 residues. On the other hand, there are significant hydrogen bond and electrostatic interaction of the oxygen atom in an ethoxyethyl group with hydroxyl group of SER 268, resulting in higher Hill inhibition than those with the only alkyl ester. It is obvious that an oxygen atom in the alkyl group of ester is very important, and this can be

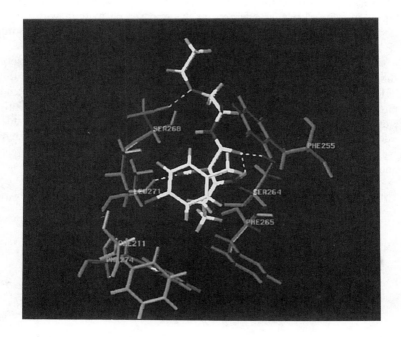

*Plate 1 Simple Complex map of Compound **1b** interacted with D1 protein of Pisum Sativum. Van der Waals and π-ring stacking interaction between it and PHE 211, PHE 255, PHE 265, PHE 274; H-bonding interaction between it and SER 264, SER 268, LEU 271.*

(Reproduced with permission from reference 9. Copyright 1999.)

(This plate is printed in color in the color insert between pages 246 and 247.)

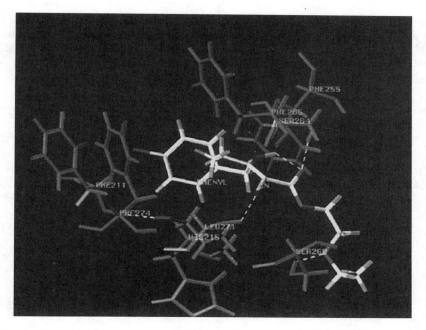

*Plate 2 Simple Complex map of Compound **1c** interacted with D1 protein of Pisum Sativum. Van der Waals and π-ring stacking interaction between it and PHE 211, PHE 255, PHE 265, PHE 274; H-bonding interaction between it and SER 264, SER 268, LEU 271.*

(Reproduced with permission from reference 9. Copyright 1999.)

(This plate is printed in color in the color insert between pages 246 and 247.)

explained by the binding of an electronegative atom with SER 268 hydroxyl group.

The phenyl moiety of cyanoacrylates is oriented into the pocket composing of PHE 211, PHE 255, PHE 265 and PHE 274, by mainly Van der Waals and π-stacking interactions. The aromatic interaction is influenced by the existence of methylene groups separating a phenyl from the amino group. Compared to compound **1a**, compound **1b** has one methylene between the phenyl and the amino group, and the phenyl moiety goes into the pocket more deeply. So compound **1b** has a closer interaction with the D1 protein, and henceforth has a higher pI_{50} value (pI_{50}=8.20) than compound **1a** (pI_{50}=5.70). So cyanoacrylates with benzylamine group could have improved interaction and binding with D1 protein, and thus have high inhibition.

It could be concluded that those cyanoacrylates with an electronegative atom in alkyl of ester group and benzyl amine group have good interaction with D1 protein, and thus exhibit high inhibition.

Based on above result, the complex model of acrylamides **1c** with D1 protein was constructed(Plate 2). Acrylamide **1c** has the key structure as acrylates **1b** necessary for binding with D1 protein, such as an electronegative atom in side amide, and benzyl amino group. So it could be expected that this acrylamide should have Hill reaction inhibition.

Synthesis and Bioassay of Cyanoacrylates(amides)

The synthetic route(9,10) to the designed compounds is shown as Figure 1.

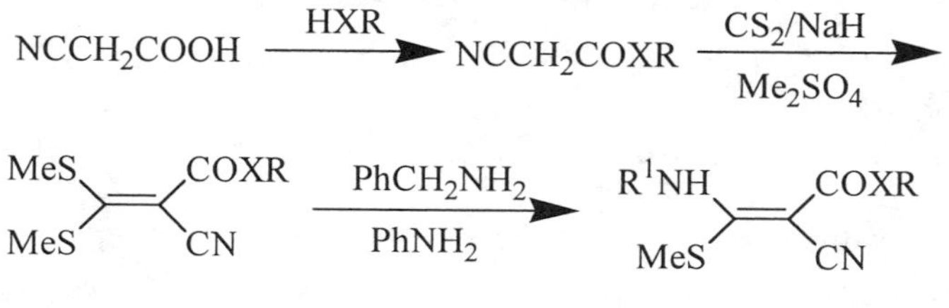

Figure 1 The synthesis route to the designed cyanoacrylates(cyanoacrylamides)

Table 2 The structures of cyanoacrylates(cyanoacrylamides), binding energies of interaction between inhibitors and D1 protein and their Hill inhibition (pI_{50} value)

$$R^1NH \diagdown \diagup COXR$$
$$MeS \diagup \diagdown CN \quad \underline{2}$$

No.	R^1	XR	pI_{50}	Binding Energy(kJ/moL)
2a	C_6H_5	OC_2H_5	5.93	−370.10
2b	$C_6H_5CH_2$	OC_2H_5	6.66	−410.73
2c	$C_6H_5CH_2$	$O\ CH_2\ CH_2O\ CH_3$	7.73	− 607.48
2d	$C_6H_5CH_2$	$O\ CH_2\ CH_2O\ CH_2\ CH_3$	8.00	− 632.94
2e	$C_6H_5CH_2$	$NH\ CH_2\ CH_2O\ CH_3$	7.42	− 564.47
2f	$C_6H_5CH_2$	$NH\ CH_2\ CH_2\ CH_2O\ CH_3$	7.28	− 483.92
2g	$C_6H_5CH_2$	OCH_2CH_2Cl	6.35	
2h	$C_6H_5CH_2$	$OCH_2CH_2OC_6H_4(CH_3)$-2	7.47	
2i	$C_6H_5CH_2$	$OCH_2CH_2OC_6H_3(CH_3)_2$-3, 4	7.20	
2j	$C_6H_5CH_2$	$OC_6H_4(OCH_3)$-4	4.33	
2k	$C_6H_5CH_2$	$OCH_2CH=CH_2$	7.10	
2l	$C_6H_5CH_2$	$OCH_2C≡CH$	6.41	

Source: Reproduced with permission from reference 9. Copyright 1999.

Compounds were assayed for Hill inhibition using suspensions of chloroplasts isolated from the leaves of 20 day old plants of *Pisum sativum*. The experimental procedure was as the method described elsewhere(11). The activity of a compound as a Hill inhibitor was expressed in terms of its pI_{50} value i. e. -lgIC$_{50}$, where IC_{50} was the molar concentration required to decrease the rate of dye reduction under illumination of saturating intensity to 50% that obtained in the absence of the compound. All the Hill inhibition of compounds **2a** ~**2l** are recorded in Table 2.

Compound **2a** and **2b** were also synthesized for comparison. Bioassay indicate that all the synthesized compounds have high Hill inhibition and compound **2a** and **2b** show lower inhibition (pI_{50}=5.93 and 6.66) as expected. Compound **2d** has higher inhibition (pI_{50}=8.00) than compound **2c**(pI_{50}=7.73), and the binding energy of compound **2d** and **2c** with D1 protein are lower than that of compound **2a** and **2b**. These results show that the binding site of D1 protein with alkyl of ester group in cyanoacrylate is large, and that the hydrophobic property might effect inhibition. On the other hand, compound **2e** and **2f** also show high Hill inhibition, and cyanoacrylamides might be a new kind of photosynthesis inhibitors. The inhibition of compound **2e** and **2f** are higher, and the binding energy of compound **2e** and **2f** with D1 protein are lower than that of compound **2a** and **2b**. This indicates that the distance between carbonyl group and electronegative atom in alkyl chain might effect inhibition.

Combinatorial Synthesis of Hydrouracil Library

As we know, various heterocyclic compounds containing nitrogen atoms, especially uracils, have biologically interesting properties in medicinal and pesticidal chemistry for example as fungicides and herbicides(12). Our studies(13) show that these uracils were lead compounds as PS II inhibitors.

Meanwhile, combinatorial chemistry has emerged as an efficient tool to synthesize "compounds libraries" for the rapid identification and optimization of new lead compounds in drug and agrochemical discovery(14). Solid-phase organic synthesis provides a rapid means for preparation of compounds libraries, and has been successfully used for the construction of oligomeric compounds and small organic molecules(15), especially the heterocyclic compounds(16) libraries. So combinatorial synthesis was introduced to optimize lead compounds uracils.

The solid-phase synthesis approach was depicted in Figure 2(17). The acryloyl chloride or acid was coupled to Wang resin **3** to afford the acryloyl ester **4**. Resin bound ester **4** was reacted with primary amines to give secondary amines **5**, which were converted to the ureas **6** by the treatment with isocyanates. Preparation of the 3-substituted hydrouracils **7** was achieved by acidic cyclization-cleavage from the resin.

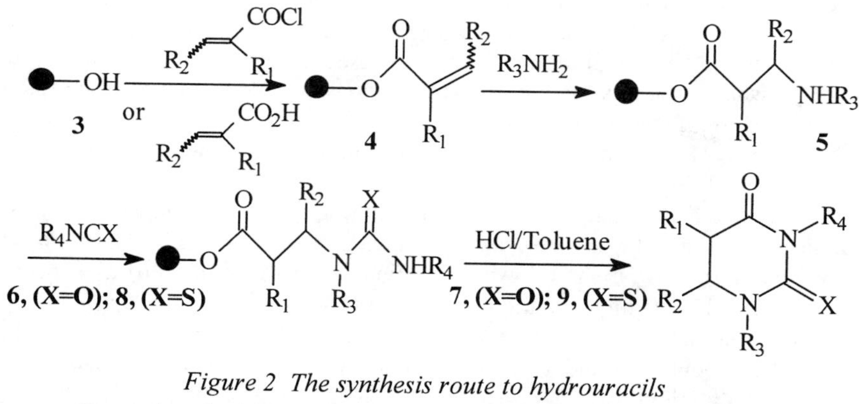

Figure 2 The synthesis route to hydrouracils
(Reproduced with permission from reference 18. Copyright 1998.)

The reaction time course was monitored using an improved KBr pellet FTIR method and MAS ^{1}HNMR method(18).

Microwave technique was introduced to improve the efficiency of solid phase organic reactions. The reaction "real time" under normal condition and under microwave irradiation are determined by analyzing the reaction time course obtained. The results indicate that the reaction process is greatly enhanced under microwave irradiation, which shows as an efficient approach for solid phase organic synthesis.

In order to improve molecular diversity of compounds library, other kinds of reagents were used as building blocks in this synthesis. Isothiocyanates were reacted with Wang resin bound amines **5** to afford thioureas **8**, which formed 2-thiohydrouracils **9** after the cleavage from the resin (Figure 2).

A small library containing 40 (2-thio)hydrouracil derivatives(R_1=H, R_2=H) was synthesized with 10 primary amines(Table 3) and 4 iso(thio)cyanates(Table 4). GC-MS analyses of the library indicated that all the desired compounds were contained in them.

The small libraries were submitted to test their photosynthesis inhibition. Using "mix and split" method(19), the active component (Figure 3, IC_{50}=8.63 ppm) was found from the library(IC_{50}=10.96ppm) with 40 compounds.

Table 3 Structure of 10 primary amines used in combinatorial libray

Table 4 Structure of 4 iso(thio)cyanates used in combinatorial libray

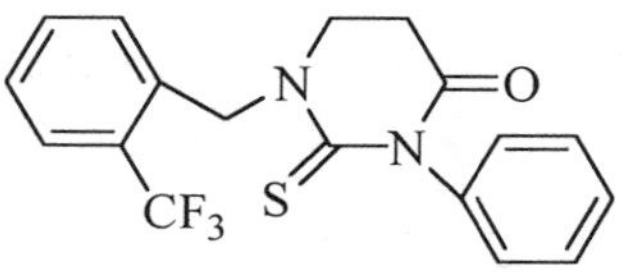

Figure 3. The structure of active component 10 from the library

Conclusions

Interaction modes of the cyanoacrylates(cyanoacrylamides) with QB binding pocket in D1 protein of *Pisum Sativum* have been presented. The essence of these modes include mainly two binding regions: phenyl moiety(Van der Waals and π-ring stacking interaction), carbonyl group and its alkyl substituent(H-bond interaction). It was found that SER 268 in D1 protein might be an important binding site. In order for compounds to exhibit strong inhibition, it is necessary to have an electronegative atom as a substituent in the alkyl ester which can undergo a H-bond interaction with SER 268 of the D1 protein. It was also found that cyanoacrylamides with the key sturcture feature of cyanoacrylates are photosynthesis inhibitors with potent activity.

We also have developed a general method for the solid-phase synthesis of (2-thio)hydrouracil analogues, which have been shown to inhibit the Hill reaction. The solid-phase approach has been applied to the combinatorial synthesis of a small library containing 40 (2-thio)hydrouracil analogs. Using "mix and split" methodology, the active component was found from the library.

Acknowledgement

This project is supported by the National Natural Science Foundation, P. R. China(29702006, 29832050), the Research Fund for the Doctoral Program of Higher Education, P. R. China(98005519), the Key Fund of Technology of Ministry of Education, P. R. China and the Key Fund of Nature Science of Tianjin, P. R. of China.

References

1. Draber, W.; Tietjen, K. G.; Kluth, J. F.; Trebst, A. *Angew. Chem. Int. Ed. Engl.* **1991**, 30, 1621-1633
2. *Pesticide Chemistry;* Frehse, H., Eds.; Advances in international research, development, and legislation; VCH Publishers Inc.: New York, NY, 1991, pp111-120
3. Sha, Y. L. Ph. D. thesis, Institute of Elemental-Organic Chemistry, Tianjin, China, 1996
4. Liu, H. Y. Ph. D. thesis, Institute of Elemental-Organic Chemistry, Tianjin, China, 1997
5. Phillips, J. N.; Huppatz, J. L. *Z. Naturforsch*, **1987**, 42C, 674-678
6. Phillips, J. N.; Huppatz, J. L. *Z. Naturforsch*, **1987**, 42C, 679-683
7. Mackay, S. P.; O'Malley, P. J. *Z. Naturforsch*, **1993**, 48C, 773-781
8. Mackay, S. P.; O'Malley, P. J. *Z. Naturforsch*, **1993**, 48C, 191-198
9. Liu,H. Y.; Sha, Y. L.; Tan, H. F.; Yang, H. Z.; Lai, L. H. *Science in China(Series B)*, **1999**, 42(3), 326-331
10. Liu, H. Y.; Lu, R. J.; Yang, H. Z. *Synthetic Communications*, **1998**, 28(21), 3965-3971
11. Tan, H. F.; Liu, H. Y.; Yang, H. Z. *Zhiwu Shengwuxu Tongxun(in Chinese)*, **1998**, 34(2), 126-129
12. Andree, R.; Drewes, M. W.; Dollinger, M. Ger. Offen DE 19, 532, 344, 1997
13. Liu, H. Y.; Yang, G. F.; Tan, H. F.; Yang, H. Z. *Chemical Journal of Chinese Universities,* **1998**, 19(12), 1946-1949
14. Terrett, N.K.; Gardner, M.; Gordon, D. W.; Kobyleck, R. J.; Steele, J. *Tetrahedron*, **1995**, 51, 8135-8173
15. Thompson, L. A.; Ellman, J. A. *Chem. Rev.* **1996,** 96, 555-600
16. Nefzi, A.; Ostresh, J. M.; Houghten, R. A. *Chem. Rev.* **1997,** 97, 449-472
17. Kolodziej, S.; Hamper, B. C. *Tetrahedron Lett.* **1996**, 37(30), 5277-5280
18. Yu, A. M.; Yang, H. Z.; Liu, H. Y.; Ma, Y.; Zhang, Z. P. *Science in China (Series B)*, **1998**, 41(5), 455-459
19. Furka, A.; Sebestyen, F.; Asgedom, M.; Dibo, G. *Int. J. Pept. Protein Res.*, 1991, 37, 487-493

Mode of Action Studies

Mode of action studies are one of the most important strategies for finding new biologically beneficial chemicals. Besides those in a narrow sense, studies on resistance and toxicology are also included in this category. From these studies, chemical structures can be modified to find paths avoiding various resistance mechanisms and to minimize the toxicity to systems other than targets as far as possible. Mode of action studies have been widely conducted and will surely become more important.

Although it was regrettable that one of the scheduled papers has been withdrawn just before the conference because of an insufficient communication between the organizing committee and the author, the remaining four papers were attractively presented in this session. They were the review for a number of new insecticidal compounds, the mode of action of a specific inhibitor of brassinosteroid biosynthesis, the three-dimensional structure-activity analysis of non-competitive antagonists at GABA receptors, and the neurotoxicology of cyclodienes in mammalian nervous systems. All of these chapters are believed to provide useful information and key steps for mode of action studies and for the designing of a variety of new agrochemicals in the future. All of which will greatly contribute to safer crop protection.

We express appreciation to Dr. David Hunt who was one of the major organizers of this section of the conference but had to retire because of a change in his work affiliation.

Keiichiro Nishimura, Ph.D.
Research Institute for Advanced Science and Technology
Osaka Prefecture University
1-2 Gakuen-cho
Sakai
Osaka 599-8570
JAPAN

Don R. Baker, Ph.D.
Berkeley Discovery
15 Muth Drive
Orinda, CA 94563

Chapter 21

New Insecticides: Mode of Action and Selective Toxicity

Robert M. Hollingworth

Center for Integrated Plant Systems and Department of Entomology,
Michigan State University, East Lansing, MI 48824

Insecticides (including acaricides) represent a pest
management technology that has gradually been improved
over 60 years to remedy, in turn, inadequate biological
spectrum and potency, excessive environmental persistence,
and more recently, high acute toxicity to vertebrates. A new
generation of chemistry is now reaching the market which
achieves high levels of safety to vertebrates through one or
more of the following (i) actions on novel, insect-specific
sites, (ii) insect-specific actions on non-specific biochemical
targets (iii) favorable pharmacokinetics, particularly through
the development of propesticides that require metabolic
activation. An overview of these new compounds with
emphasis on their mechanisms of action and selective toxicity
is presented.

The specifications for a modern pesticide seem almost impossible to meet.
On the one hand, the public demands high levels of safety, both to humans and
the environment, which requires low toxicity to a large range of non-target
species (even to other insects that are beneficial or endangered). Short
environmental persistence, and very low residues in food and water are also

required. On the other hand, growers want pesticides that have efficacy across a range of pests including resistant populations, reasonable environmental persistence to maintain control for an adequate period of time, low cost, and ease of use. High levels of safety are also desirable from the grower viewpoint to assure worker safety and minimize environmental impacts. Despite this daunting list of requirements, some of which are conflicting, a considerable number of new insecticides and acaricides which come close to satisfying these provisions have recently been developed and are entering the market. This is fortunate, because the 1996 Food Quality Protection Act will eventually cause the loss in the United States of a number of critical uses of current insecticides and replacements will be needed. In the following discussion the focus is on areas of new chemistry to replace older classes. A few organophosphates and pyrethroids are still under development, but are not included here. Additional discussion of some of these compounds may be found in references (1) and (2).

Neurotoxic Insecticides

Like their predecessors, the chlorinated hydrocarbons, organophosphates, carbamates and pyrethroids, the majority of the newer insecticides are neurotoxic. Several new classes and compounds are now available.

Neonicotinoids

These are the most important new class of insecticides in the last 30 years and are discussed in detail in a recent book (3). The forerunner of the class, imidacloprid, is used very widely around the world. A number of related compounds such as acetamiprid, nitenpyram, thiacloprid and thiamethoxam are under development in different parts of the world (Figure 1). Their utility is enhanced by their strong systemic activity in plants and they are particularly active against sucking insects. Sublethal effects such as feeding disruption are a notable aspect of their ability to prevent insect damage. The development of this class can be traced back to the minor nitromethylene insecticide nithiazine (4). These compounds bind to the acetylcholine recognition site of the nicotinic acetylcholine receptor. Such receptors are vital in vertebrates also, but the neonicotinoids bind much more avidly to insect than vertebrate receptors (3,5) which provides a considerable measure of safety as far as acute toxicity is concerned (Table I). In addition these compounds are hydrophilic and tend to be particularly safe to aquatic species, presumably due to limited uptake as well as low activity on vertebrate receptors.

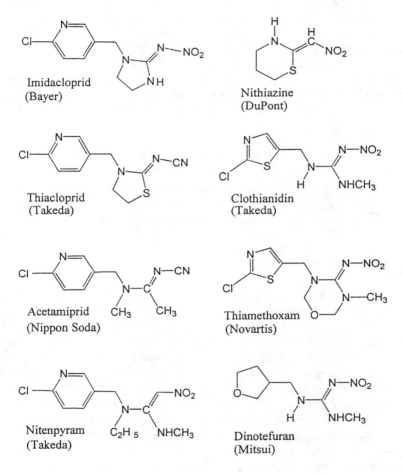

Figure 1. Structures of neonicotinoid insecticides.

Table I: Toxicity of New Insecticides to Several Non-Target Species

Compound	Acute Oral LD50 (mg/kg)		LC50 (ppm)	
	Rat	Bird[a]	Fish[b]	Daphnia
Neurotoxicants				
Acetamiprid	146-217	180	>100	>1000
Imidacloprid	450	31-152	211	85
Nitenpyram	1575-1680	1124->2250	>10->1000	>10,000
Thiamethoxam	1563	576-1552	>114	>100
Spinosyn A	>3600	>2000	96	96
Emamectin benzoate	70	--	174-180	0.99
Milbemectin	456-762	>2250	0.45	0.42
Fipronil	100	31->2150	0.34	0.19
Ethiprole	>2000	--	--	--
Bifenazate	>5000	1142	0.58-0.76	0.50
Indoxacarb	>5000	>2250	>0.5	--
Pymetrozine	5820	>2000	>100	> 100
Insect Growth Regulators				
Pyriproxifen	3773-4733	>2000	0.45-2.7	0.40
Lufenuron	>2000	>2000	>29->73	--
Novaluron	>5000	>2000	>1.0	0.058
Etoxazole	>5000	>2000	0.89->40	>40
Buprofezin	2198-2355	--	2.7	51
Methoxyfenozide	>5000	--	>4.3	3.7
Tebufenozide	>5000	>2150	3.0-5.7	3.8
Oxidative Phosphorylation Disrupters				
Fenazaquin	130-140	1750->2000	0.004-0.034	0.004
Pyridaben	820-1350	>2250	0.001-0.008	0.0006
Tebufenpyrad	600-100	>2000	0.073	1.2
Chlorfenapyr	440-1150	10-34	0.007-0.5	0.006
Fluazinam	>5000	1780-4190	0.11-0.15	0.22

[a] Bird = quail, duck or pheasant
[b] Fish = bluegill, carp or trout

Spinosyns

The spinosyns are a family of natural macrolides produced by a rare actinomycete, *Saccharopolyspora spinosa*. Their discovery and development have been reviewed (6,7). The commercial insecticide, spinosad, is derived from the fermentation broth and contains a mixture of spinosyns A and D (85:15).

Further structural modifications have shown that a synthetic analog in which ethoxy groups replace the 2',3',4'-trimethoxy groups on the rhamnose sugar of spinosyn A has considerably improved activity by ingestion or contact (6), probably due to its higher lipophilicity. Spinosad is particularly effective against lepidopteran species with a toxicity comparable to that of the pyrethroids, but the spectrum of insecticidal activity is quite broad.

The exact site of action for these compounds is not known, but it has been shown (8,9) that they activate nicotinic acetylcholine receptors in the insect central nervous system possibly by an allosteric action since they do not bind at the acetylcholine recognition site. They also interact with GABA-gated chloride channels, but this is believed to be of secondary importance in their toxicity to insects (9). Remarkably, these compounds seem to be virtually without effect on vertebrate acetylcholine receptors and have an extremely favorable spectrum of safety to vertebrates (Table I).

Emamectin Benzoate and Milbemectin

The avermectins and milbemycins, isolated from cultures of *Streptomyces* species, and their semi-synthetic analogs have been used for a number of years as anthelmintics, insecticides and acaricides (10,11,12). Abamectin (avermectin B_1) has been available in the US for some time while milbemectin (a mixture of milbemycins A_3 and A_4) is only now being introduced here. Both are primarily acaricides with some insecticidal uses. Emamectin benzoate, a semi-synthetic derivative of abamectin in which the 4"-hydroxy group is replaced with an epi-methylamino moiety is also now being registered in the US. This relatively small structural modification confers a remarkable increase in activity against lepidopteran insects (10,11). The primary mechanism of action of these compounds in insects involves an increase in the activity of inhibitory glutamate-gated chloride channels through interaction at an allosteric site (10,11,13). Stimulation of this channel leads to decreased activity and paralysis. The avermectins also interact with other chloride channels including those gated by GABA, and the stimulation of GABA-gated channels may also contribute to insecticidal effects (11). The intrinsic acute toxicity of emamectin benzoate to vertebrates is quite high (Table I), but its extreme potency against insects and consequent low use rates (8-16 g/ha) provides a good margin of safety in practice.

Fiproles

The first member of the fiprole (phenylpyrazole) insecticides, fipronil, is now a well established compound with a broad range of uses for foliar, soil or seed treatments, and for the control of ectoparasites. A second member of the series, ethiprole, is now being commercialized (Figure 2). These compounds are blockers of the GABA-gated chloride channel (14,15). Radioligand binding studies show that fipronil binds with higher affinity to insect than vertebrate GABA receptors and that the binding sites of a range of channel blockers differ in properties between insects and vertebrates (14,15). Despite this, the acute toxicity of fipronil to vertebrates is relatively high although that of ethiprole is much lower (Table I). A metabolic component is also involved in making fipronil a relatively selective insecticide (15). Differences in binding sites between different insecticide classes at this channel also explain why cross-resistance between fiproles and the cyclodiene insecticides, which also act as chloride channel blockers, is limited (14).

Bifenazate

This hydrazine derivative (Figure 2) is a new broad spectrum specific acaricide from a novel structural class. Its mechanism of activity is not known with certainty, but it has been described as having a unique action on GABA receptors in insects (16). The relevance of this observation to its specific toxicity to mites remains to be established. Because of their size, mites remain a difficult subject for mechanism of action studies. Its safety to non-target species is excellent (Table I), but the reasons for this are not yet clear.

Indoxacarb

Indoxacarb is the product of a long process of structural optimization based on the pyrazoline insecticides initially explored by Phillips-Duphar in the 1970s (17). The commercial insecticide (Figure 2) is a mixture of the S and R isomers, although the S isomer provides most of the insecticidal activity. Indoxacarb shows strong lepidopteran toxicity and is also active on some other insects. Although it attacks voltage-dependent sodium channels, which are also the target of DDT and pyrethroids, indoxacarb does so by a novel mechanism which involves a local anesthetic-like action within the ion channel resulting in a block in excitability (2,18). The work of Wing et al. (18) shows that indoxacarb is a propesticide in insects which requires activation in vivo through N-decarboxymethylation by esterase or amidase action. On the other hand,

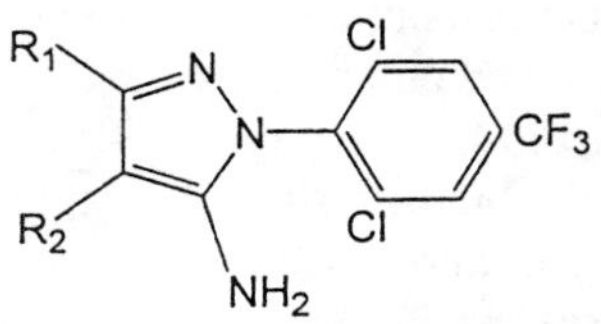

244

R₁ = H, R₂ = S(O)CF₃: Fipronil (Rhone-Poulenc)
R₁ = CN, R₂ = S(O)C₂H₅: Ethiprole (Rhone-Poulenc)

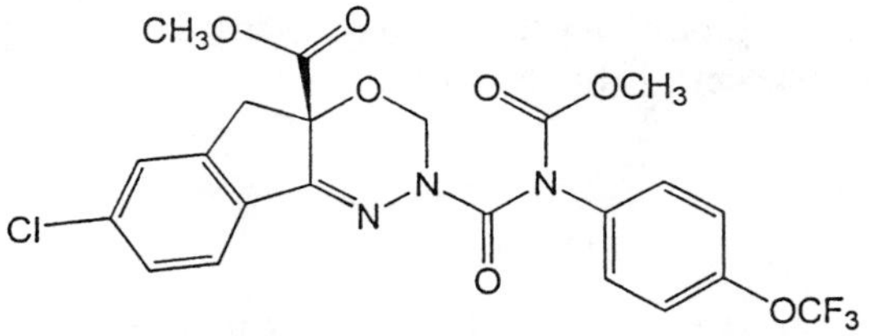

Indoxacarb (DuPont)

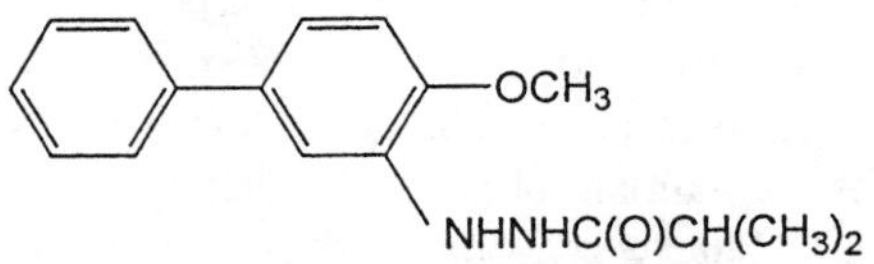

Bifenazate (Uniroyal)

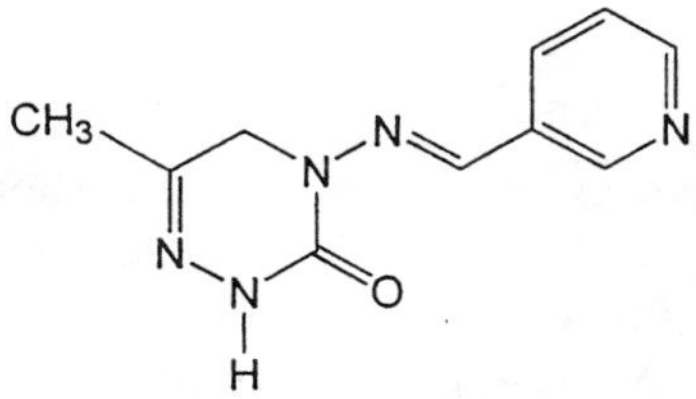

Pymetrozine (Novartis) 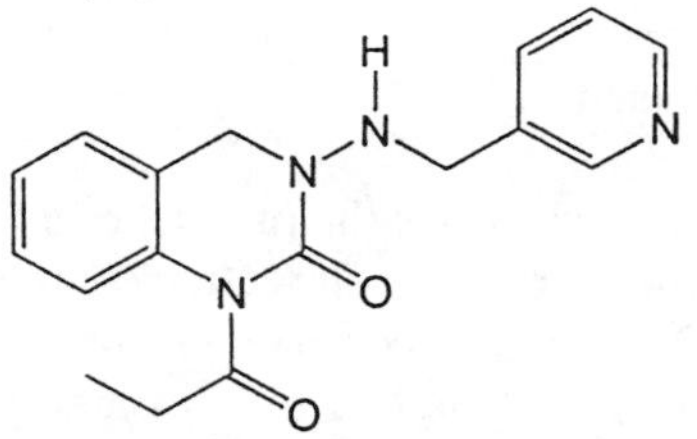   Nihon Nohyaku R768

Figure 2. Structures of some new insecticides with neurotoxic actions.

Zhao *et al.* (*19*), using a mammalian preparation, concluded that the parent compound (but not its metabolite) had potent modulatory effects on neuronal nicotinic acetylcholine receptors which were probably important in its acute toxicity. Both relative specificity at the insect sodium channel and more rapid activation in insects may underlie indoxacarb's high level of vertebrate selectivity (Table I).

Pymetrozine

This new systemic insecticide is active against sucking insects and has a novel mechanism of toxicity which involves a specific disruption of feeding behavior. Exposed aphids remove their stylets from the plant and cease feeding while retaining normal locomotion, eventually dying of starvation (*20*). A new compound, Nihon Nohyaku R-768, with structural similarities to pymetrozine, which acts by feeding disruption and with specificity for sucking insects, was announced at the 1998 IUPAC Congress of Pesticide Chemistry (*21*; Figure 2). The biochemical mechanism of action of these compounds has not been disclosed, but the very low acute toxicity of pymetrozine to vertebrates is striking (Table I).

Insect Growth Regulators

Insecticides in this class disrupt a specific stage in the growth and development of insects. Such compounds include juvenile hormone mimics, the most recent of which is pyriproxifen (now receiving registration for a variety of agricultural uses in the US), mimics of the insect molting hormone, 20-hydroxyecdysone (20-HE), and compounds that interfere with the synthesis of the new cuticle during the molting cycle (*22*). Since these processes and the hormonal control of development in insects differ significantly from those of vertebrates, such insect growth regulators tend to have a high level of vertebrate safety (Table I).

Chitin Synthesis Inhibitors (CSIs)

The original CSIs were benzoylphenylureas, typified by diflubenzuron, the first member of the group. Several members of this class are already widely used in many parts of the world, though use in the US is more limited (*22,23*). New members of the group continue to be developed. Two such compounds are lufenuron and novaluron (Figure 3). The benzoylphenylureas inhibit chitin biosynthesis in the cuticle at a site that has not been specifically identified but which probably involves interference with the proteins essential for the addition of N-acetylglucosamine units onto the growing chitin polymer chain (23,24). For this reason toxicity is limited to life stages that involve cuticular synthesis i.e. the embryonic, larval and pupal stages, and lethal effects on adults are not seen.

Although little seems to have been published to confirm the claim, a second group of specific acaricides probably act in the same way as the benzoylphenylureas. These compounds are characterized by the presence of two phenyl groups separated by a heterocyclic ring and include an older compound, clofentezine, and two recent ones, flufenzine and etoxazole (Figure 3). Evidence supporting an action similar to that of the benzoylphenylureas include structural similarities since they overlay the structure of the benzoylphenylureas quite closely, the comparability of the preferred benzoyl ring substituents between these two classes, and the fact that these acaricides also have activity against eggs and larvae, but not adults. The reason for the acaricidal specificity of these compounds is unknown. Like the benzoylphenylureas, they have excellent safety to vertebrates, but some CSIs can present a hazard to aquatic invertebrates (Table I).

Buprofezin (Figure 4) is another insecticide that inhibits chitin biosynthesis and prevents successful molting, in this case by a mechanism that may involve extending the presence of the molting hormone, 20-HE, during the later stages of the molt (25). It was developed nearly 20 years ago and has been widely used in many countries to control homopterans and mites, but it is yet another compound that is only now entering the US market.

Dibenzoylhydrazines

These insecticides (the "fenozides") represent a new class with a novel mechanism of action and have recently been reviewed (26). 20-Hydroxyecdysone is the hormone which initiates the molt in insects. The dibenzoylhydrazines mimic 20-HE in this effect which causes a premature molt to be initiated in larvae. Their ecdysonergic effects persist into the later stages of the molt when ecdysone-like activity should be declining and this disrupts further development (26,27). The initial fenozide was tebufenozide (Figure 4). More recent examples include methoxyfenozide, which is often several-fold more active than tebufenozide, halofenozide, a compound mainly directed at soil insects, and chromafenozide. The dibenzoylhydrazines have a second, neurotoxic action involving their ability to block potassium channels (28)

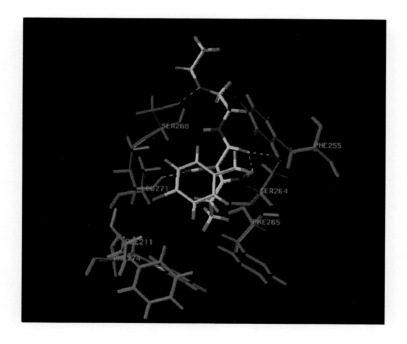

*Plate 1 Simple Complex map of Compound **1b** interacted with D1 protein of Pisum Sativum. Van der Waals and π-ring stacking interaction between it and PHE 211, PHE 255, PHE 265, PHE 274; H-bonding interaction between it and SER 264, SER 268, LEU 271.*

(Reproduced with permission from reference 9. Copyright 1999.)

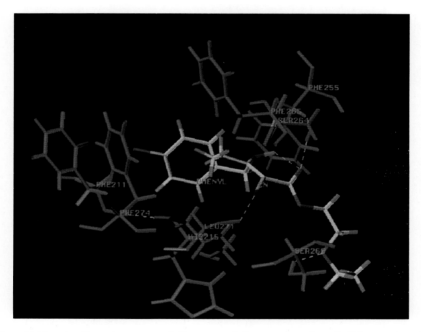

*Plate 2 Simple Complex map of Compound **1c** interacted with D1 protein of Pisum Sativum. Van der Waals and π-ring stacking interaction between it and PHE 211, PHE 255, PHE 265, PHE 274; H-bonding interaction between it and SER 264, SER 268, LEU 271.*

(Reproduced with permission from reference 9. Copyright 1999.)

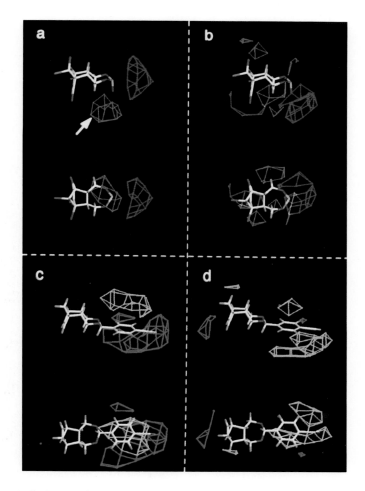

Plate 1. Orthogonal views of CoMFA field maps for non-competitive GABA antagonists. a: housefly, electrostatic; b: rat, electrostatic; c: housefly, steric; d: rat, steric.

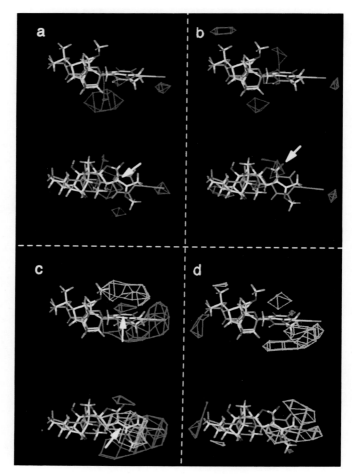

Plate 2. Orthogonal views of CoMFA field maps for picrodendrin GABA antagonists. a: housefly, electrostatic; b: rat, electrostatic; c: housefly, steric; d: rat, steric.

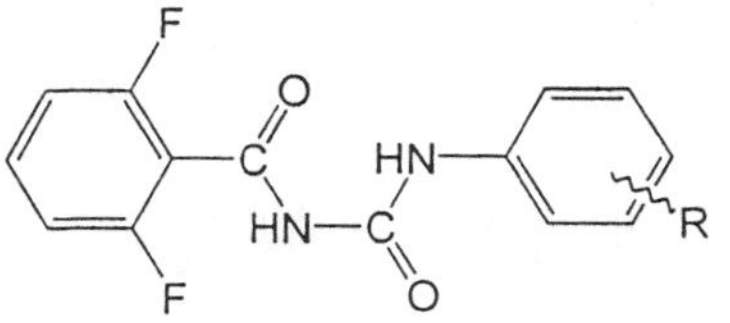

R = 4-Cl: Dimilin (Uniroyal)

R = 3-Cl, 4-OCF$_2$CHFOCF$_3$:   Novaluron (Makhteshim-Agan)

R = 2,3-diCl, 4-OCF$_2$CHFCF$_3$:   Lufenuron (Novartis)

Benzoylphenylurea Insecticides

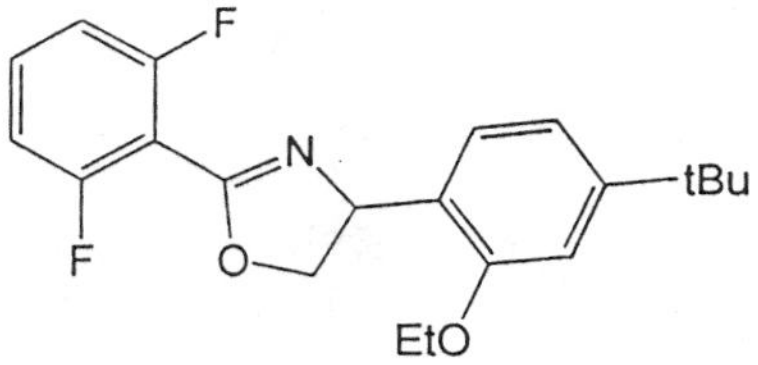

Etoxazole (Valent)

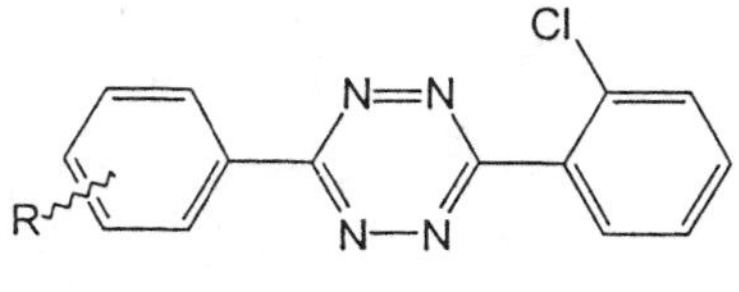

R = 2-Cl: Clofentezine (Agrevo)

R = 2,6-diF: Flufenzine (Chinoin)

Figure 3. Structures of insecticidal chitin biosynthesis inhibitors.

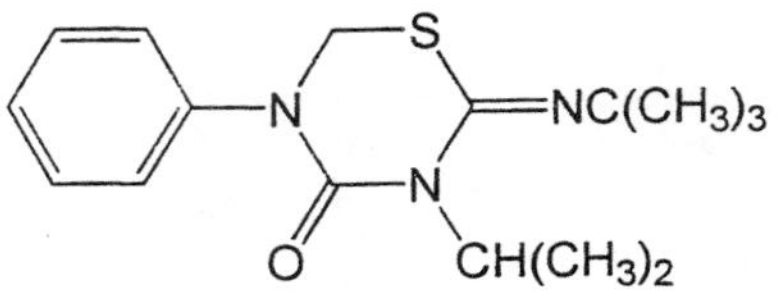

Buprofezin (Aventis)

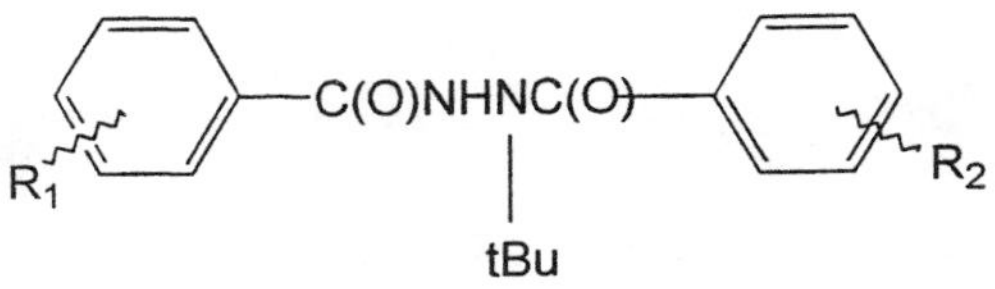

R$_1$ = 4-ethyl   Tebufenozide
R$_2$ = 3,5-dimethyl   (Rohm & Haas)

R$_1$ = 2-methyl, 3-methoxy   Methoxyfenozide
R$_2$ = 3,5-dimethyl   (Rohm & Haas)

R$_1$ = 4-chloro   Halofenozide
R$_2$ = H   (R&H/American Cyanamid)

R$_1$ = 2-methyl, 3,4-(CH$_2$CH$_2$O)   Chromafenozide
R$_2$ = 3,5-dimethyl   (Sankyo)

Figure 4. Structures of insecticides affecting insect molting through ecdysonergic actions.

which may be significant in some insects. Their safety to vertebrates is uniformly excellent (Table I).

Compounds Acting on Oxidative Phosphorylation

Inhibitors of Mitochondrial Complex I

A series of lipophilic nitrogen heterocycles are useful as acaricides by inhibiting mitochondrial respiration at complex I through interaction at the rotenone binding site. Limited insecticidal activity is also seen. These compounds include fenazaquin, fenpyroximate, pyridaben and tebufenpyrad. Their general characteristics have already been reviewed (29). Recently a new member of this class, tolfenpyrad (Figure 5), was reported which has the notable property of good activity against both lepidoptera and sucking insects (30,31). The available data suggest that these compounds do not have much target site selectivity between insects and vertebrates, and any selectivity is presumably based on pharmacokinetic differences, particularly metabolism. Most of these compounds show relatively high toxicity to aquatic species (Table I) which may be attributable to the rapid uptake of lipophilic compounds from aquatic environments, the lack of target site selectivity, and the limited metabolic rates for xenobiotics in some aquatic species.

Inhibitors of Mitochondrial Complex III

Few insecticides are known to inhibit mitochondrial respiration at complex III, so it is interesting to note the commercialization of the 2-alkyl naphthoquinone derivative, acequinocyl, first discovered at DuPont, as a specific acaricide in Japan (32; Figure 5). Acequinocyl is a propesticide requiring esterase hydrolysis to release the active hydroxynaphthoquinone which then inhibits respiration at complex III (33). Very similar 2-alkyl naphthoquinones were isolated from a Chilean plant, *Calceolaria andina,* by Khambay *et al.* Their derivatives show insecticidal activity against sucking insects (34), inhibit respiration at complex III (35), and are currently under evaluation for commercialization.

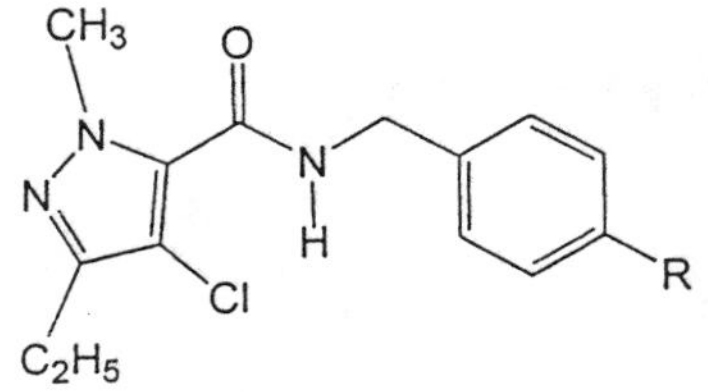

R = *t*-Bu: Tebufenpyrad (Mitsubishi)
R = *p*-Tolyloxy: Tolfenpyrad (Mitsubishi)

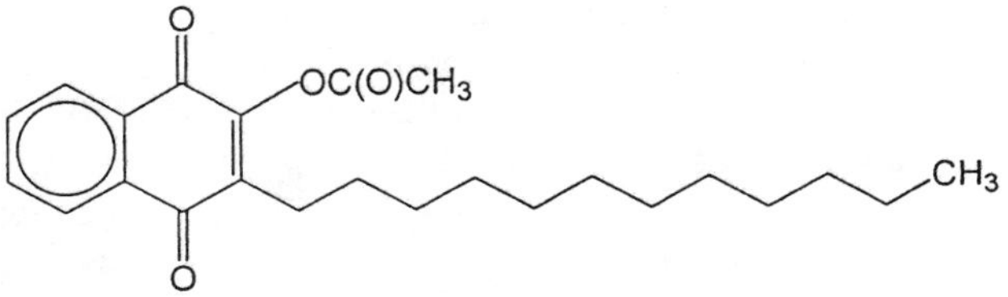

Acequinocyl (Agro Kanesho)

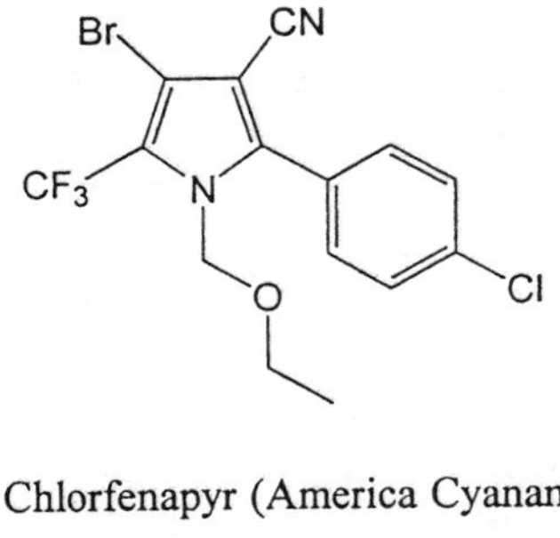

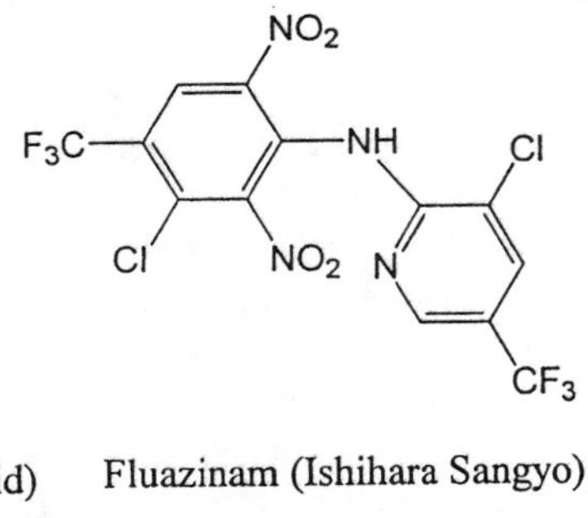

Chlorfenapyr (America Cyanamid)   Fluazinam (Ishihara Sangyo)

Figure 5. Structures of insecticides acting on oxidative phosphorylation.

Mitochondrial Uncouplers

Perhaps the greatest challenge in pesticide discovery is that of developing mitochondrial uncouplers with adequate margins of safety to non-target species. Since uncouplers are lipophilic weak acids that act as proton shuttles across the mitochondrial inner membrane, their action is purely physicochemical in nature, does not involve a specific binding site, and offers little hope for selectivity at the site of action. Any selectivity must therefore depend on pharmacokinetic differences. In one case, the fungicide/acaricide, fluazinam, considerable selectivity has been achieved (36; Table I) by making the compound open to attack by sulfhydryl groups, including those of glutathione, through the inclusion of a labile chlorine atom in the ring (Figure 5). A second example of developing a selective uncoupler, but with only partial success, is chlorfenapyr (Figure 5), a broad spectrum insecticide with many agricultural uses. This compound was developed from the structure of a natural compound, dioxapyrrolomycin, isolated from fermentations of *Streptomyces fumanus* (37,38). Again this compound is a propesticide. The N-alkoxy group must be removed by oxidase action to release the free pyrrole (AC 303,268), a lipophilic weak acid with strong uncoupling activity, before toxicity can occur (39). For a compound that generates a powerful uncoupler, chlorfenapyr shows an unusually high level of acute safety to mammals (Table I). This is probably because of a faster rate of degradation of AC 303,268 by mammals compared to insects rather than a slower rate of activation in mammals (40). However, birds and aquatic species are less well protected and considerable concern has been expressed about chlorfenapyr's environmental persistence and potential effects on avian reproduction (41).

Some Near Misses

Finally, it is worth noting two projects that have been reported where novel biochemical target sites were impacted resulting in good insecticidal activity, but which did not ultimately lead to commercial products. Compounds designed as agonists for muscarinic acetylcholine receptors were shown to have notable insecticidal activity against mites and sucking insects by scientists at Dow Elanco (42). 2,4-Diaminopyrimidines that inhibit dihydrofolate reductase were designed at FMC. These showed strong insecticidal and acaricidal activity in the laboratory but not in the field (43). The latter represents an example of insecticidal action through the inhibition of an enzyme of intermediary metabolism which is rare among insecticides but common in current herbicides and fungicides.

Conclusions

The last decade has been a very successful one for the discovery and development of a new generation of insecticides which can meet the stringent modern requirements for biodegradability, selectivity, safety and efficacy. An overview of the compounds presented here leads to several conclusions.

1. Natural products are still a very important source of new insecticides and as models for synthetic analogs, e.g. 2-alkyl naphthoquinones, chlorfenapyr, emamectin benzoate, and spinosyns.

2. There seem to be a quite limited number of biochemical targets for effective insecticides. Many of these sites (nervous system, bioenergetics) are critical for vertebrates as well as insects. Nevertheless, very safe compounds can be found which attack these common sites with considerable insect specificity, e.g. fiproles, indoxacarb, neonicotinoids, spinosyns.

3. Where pharmacodynamic differences are lacking as a basis for selectivity, pharmacokinetic factors (generally metabolic activation or deactivation) can be helpful in providing safety to non-target species (e.g. chlorfenapyr, fluazinam, indoxacarb, many lipophilic heterocyclic complex I inhibitors), but this often has serious limits, particularly for lipophilic compounds in aquatic systems.

All of these compounds, were discovered by traditional methods involving the screening of natural product mixtures from microbes and plants, or random screening of synthetic compounds followed by lead optimization. With the exception of the juvenoids, none of the newer commercial insecticides seem to have been designed from the start to impact specific targets, particularly those of specific importance to insects which offers the highest possibility for vertebrate safety through target site selectivity.

Extensive change continues in the agrochemical industry including consolidation of companies, low profitability in many important agrochemical markets, competition from generic insecticides, and the deployment of transgenic plants with insect resistance that compete for both industry's internal R&D resources and for the largest markets for insect management products. These forces make the future for the discovery of new conventional insecticides quite uncertain. However, the revolution in the discovery process arising from high throughput synthesis and screening, both *in vivo* and *in vitro*, coupled to the profusion of target site information that will come from genomics and bioinformatics, do give cause for hope that new generations of novel and selective insecticides can be found which will meet societal and grower expectations. Even so, it would be wise to regard the current generation of new insecticides as a precious, non-renewable resource which must be deployed carefully within the context of resistance management programs.

References

1. Ishaaya, I.; Horowitz, A. R. In *Insecticides with Novel Modes of Action*; Ishaaya, I.; Degheele, D., Eds.; Springer: Berlin, 1997; pp 1-24.
2. Salgado, V. L. In *Pesticide Chemistry and Bioscience*; Brooks, G. T.; Roberts, T. R., Eds.; Spec. Publ. No. 233; Royal Society of Chemistry: Cambridge, UK, 1999; pp 236-246.
3. Yamamoto, I.; Casida, J. E. *Nicotinoid Insecticides and the Nicotinic Acetylcholine Receptor;* Springer: Tokyo, 1999.
4. Kagabu, S. In *Ref. 3.*, pp. 91-106.
5. Elbert, A.; Nauen, R.; Leicht, W. In *Insecticides with Novel Modes of Action*; Ishaaya, I.; Degheele, D., Eds.; Springer: Berlin, 1997; pp 50-73.
6. Crouse, C. D.; Sparks, T. C. *Revs. Toxicol.* **1998**, *2*, 133-146.
7. DeAmicis, C. V.; Dripps, J. E.; Hatton, C. J.; Karr, L. L. In *Phytochemicals for Pest Control*; Hedin, P. A.; Hollingworth, R. M.; Masler, E. P.; Miyamoto, J.; Thompson, D. G., Eds.; ACS Sympos. Ser. No. 658; American Chemical Society: Washington, DC, 1997; pp 144-154.
8. Salgado, V. L. *Pestic. Biochem. Physiol.* **1998**, *60*, 91-102.
9. Salgado, V. L.;Watson, G. B.; Sheets, J. J. In *Proc. 1997 Beltwide Cotton Production Conf.,* National Cotton Council: Memphis, TN, 1997; pp 1082-1086.
10. Fisher, M. H. In *Phytochemicals for Pest Control*; Hedin, P. A.; Hollingworth, R. M.; Masler, E. P.; Miyamoto, J.; Thompson, D. G., Eds.; ACS Sympos. Ser. No. 658; American Chemical Society: Washington, DC, 1997; pp 220-239.
11. Jansson, R. K.; Dybas, R. A. In *Insecticides with Novel Modes of Action;* Ishaaya, I.; Degheele, D., Eds.; Springer: Berlin, 1997; pp 152-170.
12. Kornis, G. I. In *Agrochemicals from Natural Products;* Godfrey, C. R. A., Ed.; Dekker: New York, 1995; pp 215-255.
13. Duce, I. R.; Bhandal, N. S.; Scott, R. H.; Norris, T. M. In *Molecular Action of Insecticides on Ion Channels;* Clark, J. M., Ed.; ACS Sympos. Ser. No. 591; American Chemical Society: Washington, DC, 1995; pp 251- 263.

14. Gant, D. B.; Chalmers, A. E.; Wolff, M. A.; Hoffman, H. B.; Bushey, D. F. *Revs. Toxicol.* **1998**, *2*, 147-156.
15. Hainzl, D.; Cole, L. M.; Casida, J. E. *Chem. Res. Toxicol.* **1998**, *11*, 1529-1535.
16. *Bifenazate Technical Data Sheet*, Uniroyal Chemical Co.: Middlebury, CT, 1998; p 8.
17. Mulder, R.; Wellinga, K.; van Daalen, J. J. *Naturwiss.* **1975**, *62*, 531-2.
18. Wing, K. D.; Schnee, M. E.; Sacher, M.; Connair, M. *Arch. Insect Biochem. Physiol.* **1998**, *37*, 91-103.
19. Zhao, X.; Nagata, K.; Marszalec, W.; Yeh, J. Z.; Narahashi, T. *Neurotoxicol.* **1999**, *20*, 561-570.
20. Fuog, D.; Fergusson, S. J.; Flückiger, C. In *Insecticides with Novel Modes of Action*; Ishaaya, I.; Degheele, D., Eds.; Springer: Berlin, 1997; pp 40-49.
21. Uehara, M.; Shimizu, T.; Fujioka, S.; Kimura, M.; Tsubata, K.; Seo, A. *Abstr. 9th Internat. Congr. Pestic. Chem.*, London; 1998, Abstract 1D-015.
22. Perry, A. S.; Yamamoto, I.; Ishaaya, I.; Perry, R. *Insecticides in Agriculture and the Environment;* Springer: Berlin, 1998; pp 137-148.
23. Oberlander, H.; Silhacek, D. L. In *Insecticides with Novel Modes of Action*; Ishaaya, I.; Degheele, D., Eds.; Springer: Berlin, 1997; pp 92-105.
24. Londershausen, M. *Pestic. Sci.* **1996**, *48*, 269-292.
25. De Cock, A.; Degheele, D. In *Insecticides with Novel Modes of Action*; Ishaaya, I.; Degheele, D., Eds.; Springer: Berlin, 1997; pp 74-91.
26. Dhadialla, T. S.; Carlson, G. R.; Le, D. P. *Annu. Rev. Entomol.* **1998**, *43*, 545-569.
27. Smagghe, G; Degheele, D. In *Insecticides with Novel Modes of Action*; Ishaaya, I.; Degheele, D. Eds.; Springer-Verlag: Berlin, 1997; pp 25-39.
28. Salgado, V. L. *Pestic Biochem. Physiol.* **1992**, *43*, 1-13.
29. Hollingworth, R. M.; Ahammadsahib, K. I. *Rev. Pestic. Toxicol.* **1995**, *3*, 277-301.
30. Fukuchi, T.; Yoshiya, K.; Kohyama, Y.; Okui, S.; Okada, I. *Abtsr. 9th Internat. Congr. Pestic. Chem.*, London; 1998, Abstract 1D-011.
31. Okada, I.; Okui, S.; Wada, M.; Takahashi, Y, *J. Pestic. Sci.* **1996**, *21*, 305-310.
32. Kinoshita, S.; Koura, Y.; Kariya, H.; Oosaki, N.; Watanabe, T. *Abstr. 9th Internat. Congr. Pestic. Chem.*, London; 1998, Abstract 1D-023.
33. Koura, Y.; Kinoshita, S.; Takasuka, K.; Koura, S.; Osaki, N.; Matsumoto, S.; Miyoshi, H. *J. Pestic. Sci.* **1998**, *23*, 18-21.
34. Khambay, B. P. S.; Batty, D.; Beddie, D. G.; Denholm, I.; Cahil, M. R. *Pestic. Sci.* **1997**, *50*, 291-296.

255

35. Jewess, P. J.; Khambay, B. P. S.; Chamberlain, K.; Devonshire, A. L. *Abst. 9th Internat. Congr. Pestic. Chem.*, London; 1998, Abstract 4B-024
36. Hollingworth, R. M.; Gadelhak, G. G.; *Revs. Toxicol.* **1998**, *2*, 253-266.
37. Kuhn, D. G.. In *Phytochemicals for Pest Control*; Hedin, P. A.; Hollingworth, R. M.; Masler, E. P.; Miyamoto, J.; Thompson, D. G., Eds.; ACS Sympos. Ser. No. 658; American Chemical Society: Washington, DC, 1997; pp 195-205.
38. Hunt, D. A.; Treacy, M. F. In *Insecticides with Novel Modes of Action*; Ishaaya, I.; Degheele, D., Eds.; Springer-Verlag: Berlin, 1997; pp 138-151.
39. Black, B. C.; Hollingworth, R. M.; Ahammadsahib, K. I.; Kukel, C. D.; Donovan, S. *Pestic. Biochem. Physiol.* **1994**, *50*, 115-128.
40. Hollingworth, R. M. and Ahammadsahib, K. I. *Unpublished results.*
41. Williams, W. *Sci. Amer.* **1999**, *281*, 26-30.
42. Dick, M. R.; Dripps, J. E.; Orr, N. *Pestic. Sci.* **1997**, *49*, 268-276.
43. Wierenga, J. M.; Cullen, T. G.; Dybas, J. A.; Henrie, R. N. II; Peake, C. J.; Plummer, M. J. *Pest Manag. Sci.* **2000**, *56*, 233-236.

Chapter 22

Molecular Interactions of Non-Competitive Antagonists with Ionotropic γ-Aminobutyric Acid Receptors: Studies into Species Difference

Yoshihisa Ozoe[1] and Miki Akamatsu[2]

[1]Department of Life Science and Biotechnology, Shimane University, Matsue, Shimane 690–8504, Japan
[2]Division of Environmental Science and Technology, Graduate School of Agriculture, Kyoto University, Kyoto 606–8502, Japan

On the basis of structure-activity relationship (SAR) data, we constructed a model of a binding site that can accommodate structurally diverse non-competitive antagonists of ionotropic γ-aminobutyric acid (GABA) receptors and tested the validity of the hypothesis of antagonist binding interaction, using a method of three-dimensional quantitative SAR analysis. As a result, these studies clarified the molecular mechanisms of the antagonist's steric and electrostatic interactions with the binding site, as well as topographical differences between the housefly and rat binding sites. Furthermore, the SAR-based synthesis of compounds led to the discovery of a series of insecticidal non-competitive antagonists that exhibit selectivity for housefly versus rat GABA receptors.

γ-Aminobutyric acid (GABA) is the major inhibitory neurotransmitter in the nervous system of animals. GABA is excreted from nerve terminals by depolarizing stimuli, and the released GABA binds to two types of receptors, ionotropic and metabotropic GABA receptors, to mediate inhibitory signals (1,2). The ionotropic GABA receptor is a chloride ion channel that is gated by the binding of GABA. The binding of GABA causes the rapid, transient opening of the ion channel, and as a result, chloride ions enter the postsynaptic neuron, leading to hyperpolarization and the resulting inhibition of neuronal firing. The channel has been shown to consist of five subunits, labeled α_{1-6}, β_{1-4}, γ_{1-3}, and so on, in vertebrates (3). In insects, the RDL (Resistance to Dieldrin) subunit has been identified as a member of the channel-forming subunits (4).

Non-competitive antagonists of ionotropic GABA receptors are thought to bind to a site within the channel to stabilize the closed conformation of the channel, leading to the manifestation of toxicity. To date, a variety of compounds have been reported to act as the non-competitive antagonists, including plant terpenoid toxins (e.g., picrotoxinin (5), anisatin (6), and picrodendrins (7)), insecticides (e.g., lindane (8,9), dieldrin (10), and fipronil (11)), convulsants (e.g., bicyclophosphates (12), bicycloorthocarboxylates (13), and dithianes (14)), and so on. A key question at this stage is whether or not these structurally diverse antagonists bind to the same site. Although molecular biology studies revealed the fundamental structure of receptor channels, the structure of the antagonist-binding domain remains to be elucidated. Therefore, we used three-dimensional quantitative structure-activity relationship (3D-QSAR) analysis of antagonists to address the question and to gain insight into the structure of the binding site and the binding mode of antagonists, particularly aiming to elucidate the difference(s) between two animal species.

Antagonist Binding Interaction: A Hypothesis

While the GABA antagonists include structurally diverse compounds, as described briefly above, they share some common structural features. This fact and earlier study results (15,16) allowed us to postulate that there are four major subsites within the non-competitive antagonist-binding site and that the structurally diverse antagonists interact with different subsites to exhibit the antagonist action (17-19). Figure 1 illustrates the binding site model along with three typical antagonists. Subsite A probably accepts an electronegative part of ligands, i.e., heteroatoms and possibly π electron(s). On both sides of subsite A, there might be spaces (subsites C and D) that accommodate a hydrophobic or steric part and an electronegative part of ligands, respectively. In addition, we infer the existence of one more subsite, designated as B. According to this model, compounds capable of interacting with at least two of the four subsites, such as *tert*-butylbicyclophosphorothionate (TBPS), should act as antagonists.

In particular, the electronegative part of antagonists appears to play a central role in the interaction with subsite A (*20*). Table I shows the effects of phenylthiophosphonic acid derivatives on GABA-induced chloride influx into mouse brain synaptosomes. Although the derivatives are structurally similar except for the heteroatom moiety in the ring system, their potencies are markedly different. The sulfur analogue (**1**) inhibits the chloride influx induced by even high concentrations of GABA. The oxygen analogue (**2**) inhibits the chloride influx only when induced by low concentrations of GABA. The nitrogen analogue (**3**) is almost inactive at any GABA concentration. The data imply that the electrostatic interaction between the heteroatoms of antagonists and the binding site is of critical importance for antagonism.

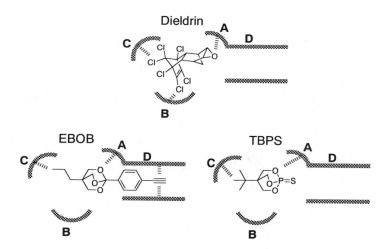

Figure 1. Proposed interactions of typical non-competitive GABA antagonists with their binding site (17-19).

Computer-aided Modeling: Revealing Species Difference

To test the hypothesis just described, we needed appropriate molecular probes. For the design of probe compounds, non-competitive antagonists were structurally divided into two groups (Figure 2). Picrotoxinin, lindane, and cyclodiene insecticides were classified into group A, and TBPS and 4'-ethynyl-4-*n*-propylbicycloorthobenzoate (EBOB) were put into group B. More than 70 antagonists were synthesized as hybrid compounds bridging the structural gap between the antagonists of the two groups (18,19). We expected that the structure-activity relationship (SAR) analysis of the hybrid compounds, dioxatricyclododecenes, would be helpful in probing the complementary binding site.

To obtain information about the structure of the binding site in rats, we determined the potency (IC$_{50}$) of a set of 64 diverse probe compounds, including dioxatricyclododecene analogues, cyclodiene analogues, lindane and its isomers, a bicycloorthobenzoate, etc., in inhibiting [^{35}S]TBPS binding (19). As an indication of potency in the binding site of the housefly, we measured the insecticidal activity (LD$_{50}$) of the same compounds (18,19). For 3D-QSAR analysis, we employed comparative molecular field analysis (CoMFA) (21), which is advantageous for the analysis of a set of bioactive compounds with different molecular skeletons. The CoMFA procedure includes: (i) the superposition of antagonist molecules according to a hypothesis; (ii) the calculation of steric and electrostatic interaction energies between the molecules and probe atoms (sp^3 carbon with a +1 charge) placed at the intersections of three-dimensional lattices surrounding the molecules; and (iii) the formulation of correlation equations between biological potencies and CoMFA field terms (22).

Table I. Effects of Phenylthiophosphonic Acid Derivatives on GABA-induced ^{36}Cl$^-$ Influx into Mouse Brain Synaptosomes

Comp. No.	Conc. (μM)	^{36}Cl$^-$ influx (% of control)			
		GABA (μM)			
		10	30	100	300
1	10	16.9	37.7	52.7	48.1
		(13.8)[a]	(10.1)	(11.7)	(3.2)
2	10	22.2	62.4	101.0	117.6
		(6.4)	(17.6)	(13.2)	(15.3)
3	10	88.7	101.8	90.2	94.1
		(13.6)	(21.4)	(9.9)	(18.0)

NOTE: [a]Standard deviation.
SOURCE: Taken from reference 20.

As a CoMFA result, we were able to obtain statistically significant correlation equations (21), supporting the view that the interacting sites of structurally diverse antagonists overlap within the antagonist-binding site. Some of the results are displayed as three-dimensional contour maps. Plate 1a and b show the results of the analysis of housefly and rat GABA receptors, respectively, and the compound shown is α-endosulfan, a cyclodiene insecticide. Red areas indicate regions where a more electronegative interaction with the binding site increases the activity, and blue areas indicate regions where a more electropositive interaction increases the activity. The blue area indicated by an arrow in Plate 1a (housefly receptors), showing an electropositive interaction, is worthy of remark with regard to species difference, because the corresponding region in the rat receptor map (Plate 1b) is rather close to a red area, showing an electronegative interaction. This suggests that the interacting amino acid residue(s) in this area might be different between both receptors.

Plate 1c shows the steric interaction of antagonists with the housefly binding site. The compound shown is a dioxatricyclododecene analogue. Green areas indicate sterically favorable regions, and yellow areas indicate sterically unfavorable regions. In this display there are yellow and green areas surrounding the benzene ring of this compound, indicating that there is a sterically favorable region below the benzene ring. In contrast, the benzene ring is surrounded by yellow contours in rat receptors (Plate 1d). It appears that there is a binding pocket that accommodates the benzene ring.

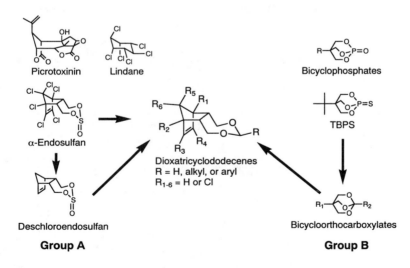

Figure 2. Structural modification of non-competitive GABA antagonists.

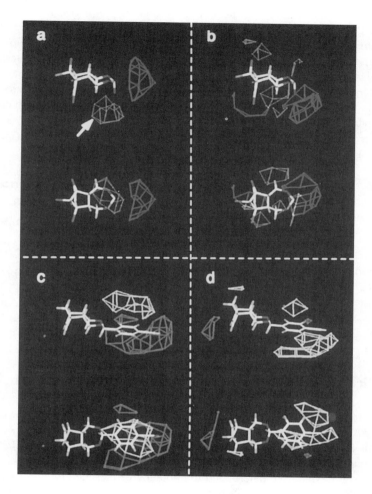

Plate 1. Orthogonal views of CoMFA field maps for non-competitive GABA antagonists. a: housefly, electrostatic; b: rat, electrostatic; c: housefly, steric; d: rat, steric.

(This plate is printed in color in the color insert between pages 246 and 247.)

Based on the CoMFA results, we were thus able to clarify the major differences in the structure of the antagonist-binding site between housefly and rat GABA receptors (Figure 3). One of the differences was found to be between subsites A and B (see Figure 1 for the location of subsites). The electropositive substituent of antagonists might be favorable in the interaction with this region of the housefly binding site, whereas the electronegative substituent might be favorable in the rat binding site. Another difference is related to the size of the space near subsite D. The space in housefly receptors is probably wider than that in rat receptors.

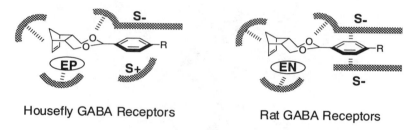

Housefly GABA Receptors   Rat GABA Receptors

Figure 3. Presumed differences in the structure of the antagonist-binding site between houseflies and rats. S+: Sterically favorable region. S-: Sterically unfavorable region. EN: Electronegative moiety. EP: Electropositive moiety.

Picrodendrins: Naturally Occurring Antagonist Probes

Picrodendrins are a series of terpenoids isolated from the Euphorbiaceae plant, *Picrodendron baccatum* (L.) Krug and Urban. We have recently shown that this series of terpenoids are non-competitive GABA antagonists (7,23,24), and we have used them as probes to test the validity of the computer-generated binding site model. We determined the potency of picrodendrins in inhibiting [³H]EBOB binding to housefly head and rat brain membranes. The relation between picrodendrins' structures and pIC_{50} (-log IC_{50} (M)) values was analyzed using the traditional Hansch-Fujita and CoMFA methods (24).

Major picrodendrins can be divided into two groups, analogues bearing a double bond between C-16 and C-18 and analogues bearing a single bond at the same position (Figure 4). While the double-bond analogues exhibit the same level of potency toward housefly and rat GABA receptors, the single-bond analogues show selectivity for housefly versus rat GABA receptors (24). The equation obtained by the Hansch-Fujita analysis of housefly receptor-binding data is shown in Figure 4. In this equation, $q(C^{16})$ is a charge on C-16, and I_{4OH} and I_{8OH} are indicator variables for the presence or absence of the hydroxyl group at either the C-4 or C-8 position. This equation indicates that a more negative charge on C-16 increases the potency of picrodendrins, and that the hydroxyl group at the C-4 or C-8 position decreases the potency. In the case of rat receptors, the number of active picrodendrins was not sufficient for us to perform a detailed analysis. However, potent picrodendrins in rat receptors proved to have more negative charges on the 17-carbonyl oxygen atom.

Plate 2a shows the orthogonal view of the field-fit overlay of picrodendrin Q (Figure 4, left, R_1, R_2 = H, R_3 = CH_3, R_4 = OCH_3) on the electrostatic potential CoMFA contour map of housefly receptors, along with EBOB (orange). The electronegative electrostatic-potential (red) area appears near C-16 of picrodendrin Q, which is shown by an arrow, in accord with the result of the Hansch-Fujita analysis, indicating that this interaction is important for high activity in housefly receptors. Plate 2b is a similar view for rat receptors. In this case, the electronegative electrostatic-potential area appears near the 17-carbonyl oxygen atom of picrodendrin Q, as indicated by an arrow. This finding is consistent with the qualitative SAR of picrodendrins described above.

Plate 2c depicts the steric map for housefly receptors. As indicated by arrows, picrodendrin Q has a double bond with planar substituents, which are accepted by the sterically favorable region. The wide green area appears to accept the corresponding nonplanar substituents of single-bond analogues as well (map not shown). Plate 2d depicts the steric map for rat receptors. In this case, this binding pocket can accommodate analogues bearing a double bond, but analogues bearing a single bond cannot be fit into the binding pocket because their nonplanar substituents protrude into the sterically unfavorable (yellow) region. This may be one of the reasons why single-bond analogues exhibit selectivity for housefly receptors versus rat receptors.

$$pIC_{50} \text{ (EBOB, fly)} = -2.87\ q(C^{16}) - 1.42\ (I_{4OH} + I_{8OH}) + 6.93$$
$$n = 12,\ s = 0.43,\ r = 0.88$$

Figure 4. Structure-activity relationships of picrodendrin GABA antagonists.

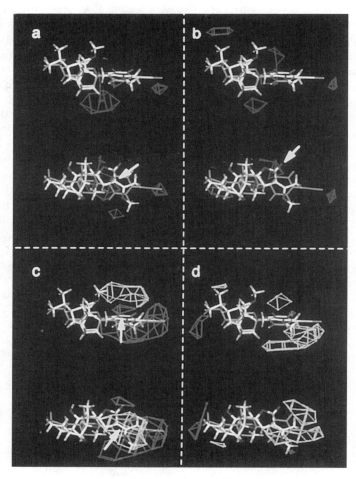

Plate 2. Orthogonal views of CoMFA field maps for picrodendrin GABA antagonists. a: housefly, electrostatic; b: rat, electrostatic; c: housefly, steric; d: rat, steric.

(This plate is printed in color in the color insert between pages 246 and 247.)

Thus, our binding site model explains the SAR of picrodendrin antagonists well. In other words, the results support the validity of our binding site model.

Bicyclophosphorothionate Antagonists Selective for Insect GABA Receptors

Fipronil, a recently developed insecticide, and related heterocycles are non-competitive antagonists exhibiting selectivity for insect GABA receptors (25,26). Insect receptor-selective antagonists thus serve as practical insecticides. Utilizing the information about the structure of the antagonist-binding site obtained by the CoMFA study, we can design antagonist compounds selective for insect GABA receptors.

As shown in Table II, bicyclophosphorus esters with a *tert*-butyl group at the 4-position are highly toxic to mammals (27) and are potent inhibitors of [³H]EBOB binding to rat GABA receptors (28). However, the introduction of an isopropyl group into the 3-position and the replacement of the 4-*tert*-butyl group with an isobutyl group reduce the activity in rat receptors and increase the activity in housefly receptors (28). This structural change was also associated with an enhancement of insecticidal activity. The 3-isopropyl group should interact with a region characteristic of housefly receptors, which was revealed to exist between subsites A and B by the CoMFA study (Plate 1a).

The effect of the isopropyl group on receptor selectivity was also observed when the bridgehead substituent was varied. Bicyclophosphorothionates bearing an *n*-propyl or an *n*-butyl group at the 4-position, in combination with the 3-isopropyl group, were also more potent in housefly receptors than in rat receptors. These compounds are selective antagonists for housefly GABA receptors; their receptor selectivity, estimated from the IC_{50} ratio (IC_{50} rat/IC_{50} housefly) in the [³H]EBOB-binding assay, being more than 50 (28).

Conclusions

The data from our laboratories point to the possibility that structurally diverse groups of naturally occurring and synthetic GABA antagonists interact with different subsites within the same binding site. On the other hand, other studies suggest that there are several distinct binding sites for structurally different antagonists, such as fipronil, cyclodienes, TBPS, or EBOB (11,29,30). Further studies would be needed to clarify this matter. We also reveal binding-site topologies based on 3D-QSAR and show that the structure of the binding site might be different between housefly and rat GABA receptors. The 3D-

QSAR results should offer opportunities for the rational design of selective antagonists for insect GABA receptors and safe insecticides.

Table II. Bicyclophosphorus GABA Antagonists: Toxicity and Potency in Inhibiting [³H]EBOB Binding

R_1	R_2	X	$LD_{50}{}^a$		$IC_{50}{}^b$	
			Housefly[c]	Mouse[d]	Housefly	Rat
t-C$_4$H$_9$	H	O	0.12[e]	0.053	nd[f]	nd
i-C$_4$H$_9$	H	O	0.17[e]	0.24	nd	nd
t-C$_4$H$_9$	H	S	1.7[g]	nd	1.6	0.062
i-C$_4$H$_9$	i-C$_3$H$_7$	S	0.049[g]	nd	0.045	2.4

NOTE: [a]Toxicity. [b]Inhibition of [³H]EBOB binding, μM. [c]μg/fly. [d]mg/kg. [e]Injected. [f]Not determined. [g]Topical.
SOURCE: Taken from references 27, 28, and 31.

Acknowledgments

We are grateful to Emeritus Prof. T. Fujita (Kyoto University) for helpful suggestions on QSAR. This study was supported in part by a Grant-in-Aid for Scientific Research from the Ministry of Education, Science, Sports, and Culture of Japan.

References

1. Schofield, P. R.; Darlison, M. G.; Fujita, N.; Burt, D. R.; Stephenson, F. A.; Rodriguez, H.; Rhee, L. M.; Ramachandran, J.; Reale, V.; Glencorse, T. A.; Seeburg, P. H.; Barnard, E. A. *Nature* **1987**, *328*, 221-227.
2. Kaupmann, K.; Huggel, K.; Heid, J.; Flor, P. J.; Bischoff, S.; Mickel, S. J.; McMaster, G.; Angst, C.; Bittiger, H.; Froestl, W.; Bettler, B. *Nature* **1997**, *386*, 239-246.
3. Barnard, E. A.; Skolnick, P.; Olsen, R. W.; Mohler, H.; Sieghart, W.; Biggio, G.; Braestrup, C.; Bateson, A. N.; Langer, S. Z. *Pharmacol. Rev.* **1998**, *50*, 291-313.

4. Hosie, A. M.; Aronstein, K.; Sattelle, D. B.; ffrench-Constant, R. H. *Trends Neurosci.* **1997**, *20*, 578-583.
5. Takeuchi, A.; Takeuchi, N. *J. Physiol.* **1969**, *205*, 377-391.
6. Ikeda, T.; Ozoe, Y.; Okuyama, E.; Nagata, K.; Honda, H.; Shono, T.; Narahashi, T. *Br. J. Pharmacol.* **1999**, *127*, 1567-1576.
7. Ozoe, Y.; Hasegawa, H.; Mochida, K.; Koike, K.; Suzuki, Y.; Nagahisa, M.; Ohmoto, T. *Biosci. Biotechnol. Biochem.* **1994**, *58*, 1506-1507.
8. Ghiasuddin, S. M.; Matsumura, F. *Comp. Biochem. Physiol.* **1982**, *73C*, 141-144.
9. Tokutomi, N.; Ozoe, Y.; Katayama, N.; Akaike, N. *Brain Res.* **1994**, *643*, 66-73.
10. Nagata, K.; Narahashi, T. *J. Pharmacol. Exp. Ther.* **1994**, *269*, 164-171.
11. Gant, D. B.; Chalmers, A. E.; Wolff, M. A.; Hoffman, H. B.; Bushey, D. F. *Rev. Toxicol.* **1998**, *2*, 147-156.
12. Casida, J. E.; Lawrence, L. J. *Environ. Health Perspect.* **1985**, *61*, 123-132.
13. Casida, J. E.; Palmer, C. J.; Cole L. M. *Mol. Pharmacol.* **1985**, *28*, 246-253.
14. Li, Q. X.; Casida, J. E. *Bioorg. Med. Chem.* **1994**, *2*, 1423-1434.
15. Soloway, S. B. *Adv. Pest Control Res.* **1965**, *6*, 85-126.
16. Matsumura, F.; Ghiasuddin, S. M. *J. Environ. Sci. Health* **1983**, *B18*, 1-14.
17. Ozoe, Y.; Matsumura, F. *J. Agric. Food Chem.* **1986**, *34*, 126-134.
18. Ozoe, Y.; Sawada, Y.; Mochida, K.; Nakamura, T.; Matsumura, F. *J. Agric. Food Chem.* **1990**, *38*, 1264-1268.
19. Ozoe, Y.; Takayama, T.; Sawada, Y.; Mochida, K.; Nakamura, T.; Matsumura, F. *J. Agric. Food Chem.* **1993**, *41*, 2135-2141.
20. Ozoe, Y.; Niina, K.; Matsumoto, K.; Ikeda, I.; Mochida, K.; Ogawa, C.; Matsuno, A.; Miki, M.; Yanagi, K. *Bioorg. Med. Chem.* **1998**, *6*, 73-83.
21. Akamatsu, M.; Ozoe, Y.; Ueno, T.; Fujita, T.; Mochida, K.; Nakamura, T.; Matsumura, F. *Pestic. Sci.* **1997**, *49*, 319-332.
22. Cramer, R. D., III; Patterson, D. E.; Bunce, J. D. *J. Am. Chem. Soc.* **1988**, *110*, 5959-5967.
23. Hosie, A. M.; Ozoe, Y.; Koike, K.; Ohmoto, T.; Nikaido, T.; Sattelle, D. B. *Br. J. Pharmacol.* **1996**, *119*, 1569-1576.
24. Ozoe, Y.; Akamatsu, M.; Higata, T.; Ikeda, I.; Mochida, K.; Koike, K.; Ohmoto, T.; Nikaido, T. *Bioorg. Med. Chem.* **1998**, *6*, 481-492.
25. Cole, L. M.; Nicholson, R. A.; Casida, J. E. *Pestic. Biochem. Physiol.* **1993**, *46*, 47-54.
26. Ozoe, Y.; Yagi, K.; Nakamura, M.; Akamatsu, M.; Miyake, T.; Matsumura, F. *Pestic. Biochem. Physiol.* **2000**, *66*, 92-104.
27. Eto, M.; Ozoe, Y.; Fujita, T.; Casida, J. E. *Agric. Biol. Chem.* **1976**, *40*, 2113-2115.
28. Ju, X.-L.; Ozoe, Y. *Pestic. Sci.* **1999**, *55*, 971-982.

29. Deng, Y.; Palmer, C. J.; Casida, J. E. *Pestic. Biochem. Physiol.* **1993**, *47*, 98-112.

30. Rauh J. J.; Benner, E.; Schnee, M. E.; Cordova, D.; Holyoke, C. W.; Howard, M. H.; Bai, D.; Buckingham, S. D.; Hutton, M. L.; Hamon, A.; Roush, R. T.; Sattelle, D. B. *Br. J. Pharmacol.* **1997**, *121*, 1496-1505.

31. Ozoe, Y.; Mochida, K.; Nakamura, T.; Shimizu, A.; Eto, M. *J. Pesticide Sci.* **1983**, *8*, 601-605.

Chapter 23

Mode of Action of Brassinazole: A Specific Inhibitor of Brassinosteroid Biosynthesis

Tadao Asami, Yong Ki Min, Katsuhiko Sekimata, Yukihisa Shimada, Jing Ming Wang, Shozo Fujioka, and Shigeo Yoshida

RIKEN (The Institute of Physical and Chemical Research), Wako, Saitama 351–0198, Japan

Inhibitors of phytohormone have proven to be useful tools for understanding hormonal function. Recently, brassinolide has been designated as a new class of phytohormone based on the physiological responses of brassinolide-deficient mutants. However, information on other roles of this hormone is limited because studies have been confined to mutants in a limited number of plant species. Therefore, specific inhibitors of brassinosteroid biosynthesis would be valuable tools for investigating their roles at various stages of plant development, such as germination, leaf expansion and flowering. Recent advances in developing brassinosteroid biosynthesis inhibitor, brassinazole (Brz), have shown the importance of brassino-steroids in broad aspects of plant growth and development. This inhibitor induced drastic morphological changes in treated plants, almost identical to those found in brassinosteroid-deficient mutants. The normal phenotype of inhibitor-treated plants could be recovered by the addition of brassinolide. This result suggests that brassinosteroids are essential for plant growth, and that specific brassinosteroid biosynthesis inhibitors can be used to clarify the functions of brassinosteroids in plants when used as a complement to brassinosteroid-deficient mutants. The action site of brassinazole was an oxidative processes from 6-oxo-campestanol to teasterone.

Introducution

Since the establishment of brassinosteroid chemistry in the late 1980s, many brassinosteroid homologues have been shown to cause remarkable responses in plants, including stem elongation, pollen tube growth, leaf bending, leaf unrolling, root growth inhibition, proton pump activation (1), promotion of ethylene production (2), tracheary element differentiation (3, 4), and cell elongation (5). Following the isolation and identification of putative intermediates, extensive studies on brassinosteroid biosynthesis have started to place these intermediates in their appropriate position in the biosynthesis pathway (6, 7). At present, over 40 brassinosteroids have been identified and most C28-brassinosteroids are thought to be biosynthesized from campesterol (13), a common plant sterol having the same side chain carbon skeleton as brassinolide (24). Despite the intensive studies on brassinosteroid biosynthesis, brassinosteroids were not recognized as a new class of phytohormone until the characterization of brassinosteroid-deficient mutants. By coupling molecular genetics with biosynthetic studies (8, 9), several *Arabidopsis* mutants with characteristic dwarfism have been isolated. These are known to be defective in the brassinosteroid biosynthesis pathway: e.g., *dwarf1* (*dwf1* (10); *dim* (11); *cbb1* (12)), *constitutive photomorphogenesis and dwarfism* (*cpd* (13)), and *deetiolated2* (*det2* (14, 15)). Recently a dwarf mutant of pea has also been shown to be brassinosteroid deficient (16). In all of these cases, applications of brassinolide restored normal growth in the dwarf mutants.

Role of Brassinosteroid Biosynthesis Inhibitors in Plant Physiology

Although the above findings indicate that brassinosteroids have an essential role in plant growth and development, detailed experimental strategies are required to identify their specific physiological effects at the levels of tissues, cells and at various developmental stages. In general, specific biosynthesis inhibitors have been effective in the determination of physiological functions of endogenous substances. The roles of such inhibitors in plant physiology are best illustrated by studies on gibberellin (GA) biosynthesis (17). For example, BX-112 and its derivatives, which are cyclohexanetrione-type chemicals, are potent plant growth regulators (PGRs) and inhibit primarily the 3β-hydroxylation step in GA biosynthesis (GA20 to GA1). They were used as chemical probes to examine GA biosynthesis and physiological roles of GA (18). If we could have a specific inhibitor of brassinosteroid biosynthesis in our hand, it could provide us with a new and complementary approach to understanding the functions of brassino-steroids, but until recently there has been no report concerning such inhibitors.

Preferable Target Sites for Brassinosteroid Biosynthesis Inhibitors

The sterol biosynthetic pathway in plants is characterized by specific steps, such as the cyclization of squalene oxide (3) to cycloartenol (4), from which plant sterols are synthesized through a series of reactions, including single or double methylations. The steps downstream of cycloartenol to brassinolide (24) are shown in Figures 1 and 2. Although the steps shown in Figures 1 and 2 occur in most plants, there are important species-specific differences. Most of the enzymes catalyzing the steps from squalene (2) are associated with membranes of the endoplasmic reticulum (ER); therefore inhibitors of these steps should have some affinity for ER membranes (19). Several kinds of inhibitors block the steps between cycloartenol (4) and campesterol (13) and they are known to be useful probes for investigating plant sterol function (20). The azole derivatives are one of the major classes of these inhibitors. Further studies on the azole derivatives indicated that at high concentrations these inhibitors blocked plant sterol biosynthesis at the site where obtusifoliol 14α-demethylase, a specific form of cytochrome P450 dependent monooxygenase, catalyzes reactions in both whole plants and cell cultures to produce phytotoxicity. Eventually, these inhibitors reduced the level of brassinolide, but the reduction of normal phytosterols and/or an accumulation of 14α-methylsterols lead to phytotoxicity because these phytosterols, such as sitosterol, stigmasterol and 24-methylcholesterol play important roles in regulating membrane fluidity and permeability, modulating the activity of membrane-bound enzymes and are required for cell proliferation (22). On the basis of these findings, it appears preferable that a specific inhibitor of brassinosteroid biosynthesis should target the biosynthetic pathway downstream of sterol biosynthesis, i.e., after obtusifoliol (7) 14α-demethylase. At present, the step catalyzing the conversion of episterol (10) to 5-dehydroepisterol (11), encoded by the DWF7 gene of *Arabidopsis*, has been shown to be the most upstream step leading to brassinosteroid biosynthesis (23). Therefore, it is probable that there are more receptive inhibitor targets downstream of this step that would have less of an intrinsic phytotoxicity.

Basis for the Design of Brassinosteroid Biosynthesis Inhibitor

Various triazole compounds are known to inhibit many cytochrome P450s in the following manner: 1) the basic nitrogen of triazole ring is positioned toward the central iron of porphyrin system and 2) the hydrophobic moiety of triazole inhibitor interact with the site normally occupied by the substrate. Thus the binding of the oxygen molecule that would normally be activated and transferred

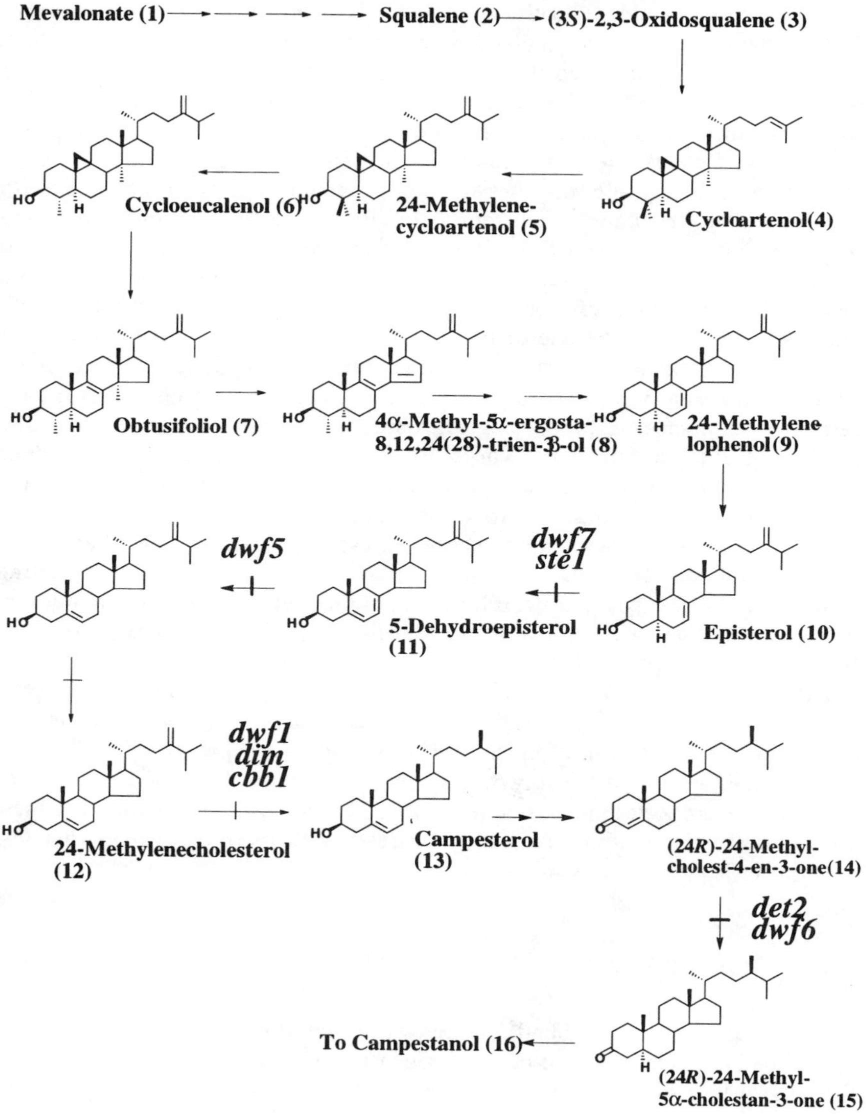

272

Figure 1. Brassinosteroid biosynthesis pathway in plants I. Brassinosteroid-deficient mutants are indicated in bold-italic.

273

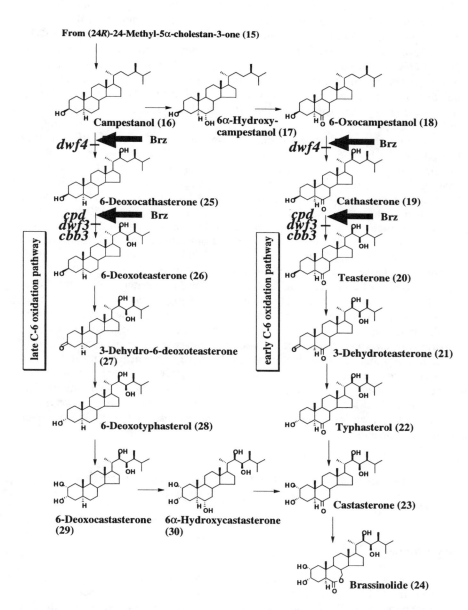

Figure 2. Brassinosteroid biosynthesis pathway in plants II. The presumptive target sites of brassinosterod biosynthesis inhibitor are indicated by bold arrow. Brassinosteroid-deficient mutants are indicated in bold-italic.

to the substrate was blocked (24). Many steps of brassinosteroid biosynthesis, after episterol (10), are catalyzed by cytochrome P450 enzymes, including the conversion of campestanol (16) to 6α-hydroxycampestanol (17), 6-oxocampestanol (18) to cathasterone (19), cathasterone (19) to teasterone (20), typhasterol (22) to castasterone (23), and castasterone (23) to brassinolide (25) (25). It is interesting to note two reports. Yokota et al. observed a slight reduction in the concentration of endogenous castasterone when plants were treated with uniconazole, a triazole-type GA biosynthesis inhibitor (26). Iwasaki and Shibaoka reported that uniconazole inhibited brassinolide-induced tracheary element differentiation (3).

On the basis of the above information, we selected triazole derivatives as lead compounds for brassinolide biosynthesis inhibitors. We examined the possibility of modifying uniconazole and paclobutrazol (Figure 3), both typical GA biosynthesis inhibitors, to produce a specific brassinosteroid biosynthesis inhibitor by reducing its potency as an inhibitor of GA biosynthesis. Since the triazole group is thought to be essential for the binding of cytochrome P450, we tried to modify the hydophobic part of the triazole derivatives other than the triazole ring according to the idea that the structure of hydrophobic part of triazole derivatives results in the selectivity for the inhibition of particular cytochrome P450. A typical example can be seen in the structure-activity relationship between $R(-)$ and $S(+)$ optical isomers of a uniconazole analog (27). The $R(-)$ isomer is more fungitoxic than the $S(+)$ isomer, whereas $S(+)$ isomer shows higher plant growth regulatory activity than $R(-)$ isomer (28). This result suggests the importance of the structure of the hydrophobic part for the binding to specific cytochrome P450. The functions of enzymes committed with steroid biosynthesis in plant are similar to those in fungi, but substrate specificity of them are different from those in fungi. This result can explain the selectivity of $S(+)$-uniconazole, which may disturb plant development by targeting a cytochrome P450 function that is involved in brassinosteroid biosynthesis. These observations suggest that it would be beneficial to screen for a specific inhibitor of brassinosteroid biosynthesis among triazole compounds.

Finding of the Brassinosteroid Biosynthesis Inhibitor, Brassinazole (Brz)

Assay Methods

Recently, our group reported that some triazole derivatives inhibit brassinosteroid biosynthesis in a way almost totally unrelated to their effects on GA biosynthesis (29). To our knowledge this is the sole report on specific

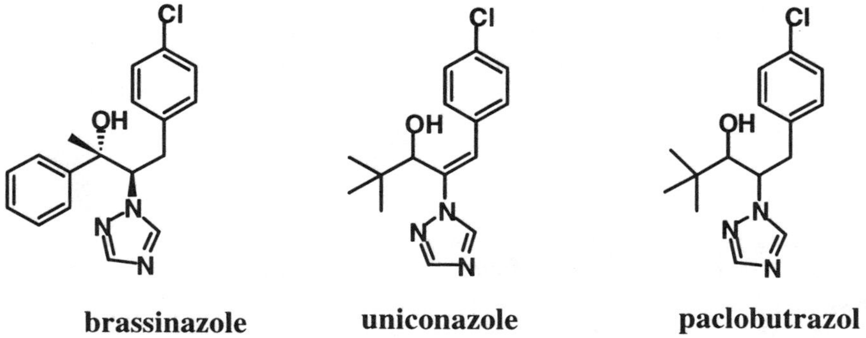

Figure 3. Structures of uniconazole, paclobutrazol and brassinazole (Brz).
Brz consists of two enantiomers, one of which is ten times more
active in inhibition of brassinosteroid biosynthesis than the other.

brassinosteroid biosynthesis inhibitors. Therefore our work was focused on, in the remainder of this report, the experimental strategy being to design and synthesize molecules based on the structure of uniconazole and paclobutrazol (Figure 3). Candidates for brassinosteroid biosynthesis inhibitors that we synthesized were subjected to biological assays. Compounds were first assayed using a rice stem elongation test to identify and eliminate GA biosynthesis inhibitors, because it is well known that these inhibitors retard rice stem elongation. Chemicals synthesized for the screening of brassinosteroid biosynthesis inhibitors were bioassayed with *Arabidopsis* and cress (*Lepidium sativum* L.) seedlings. These assay systems are particularly useful because they are sensitive to the inhibitors, and the growth inhibition could be reversed by the addition of brassinolide, a potent brassino-steroid. Brassinolide has also been shown to be effective in rescuing the *Arabidopsis* mutants *det2* and *cpd*, which show strong dwarfism with curly dark green leaves in the light and a de-etiolated phenotype with short hypocotyls and open cotyledons in darkness and are rescued by brassinolide, but not by other plant hormones such as auxins and the GAs (9). Screening for brassinosteroid biosynthesis inhibitors was performed to find chemicals which induce dwarfism in *Arabidopsis*, resembling brassinosteroid deficient mutants, which could be rescued by the addition of brassinolide. In screening experiments involving cress, which has previously been used to investigate the effects of brassinolide (30, 31), the most potent chemical compound has been identified and named brassinazole (Brz) (see the structure shown in Figure 3). Brz inhibited the growth of *Arabidopsis* and cress hypocotyls. An important observation was the recovery of *Arabidopsis* and cress growth after Brz-induced hypocotyl dwarfism by co-application of brassinolide with Brz. On the other hand, gibberellic acid was less effective than brassinolide on Brz-induced dwarfism. This implies that the morphological changes in plants treated with Brz are mainly due to a deficiency of brassinosteroids. Thus, the main target of Brz appears to be brassinosteroid biosynthesis.

Target Site(s) of Brz

To determine the biosynthetic step(s) of brassinosteroids affected by Brz, the effects of the biosynthetic intermediates downstream of cathasterone on hypocotyl elongation of inhibitor-treated plants was examined in the *Arabidopsis* and cress bioassays. We demonstrated that at least one of the target sites of Brz in brassinosteroid biosynthesis pathway is the oxidation of cathasterone to teasterone, catalyzed by *CPD* (13), a cytochrome P450 enzyme. The C22 and C23 positions of brassinosteroids are successively hydroxylated by cytochrome P450 enzymes encoded by *DWF4* (32) and *CPD* (13). The enzymes catalyzing these two steps are different, but their functions and DNA sequences are similar to each other (32). Thus, it is possible to postulate that Brz inhibits both steps. Even in our most sensitive bioassay system using *Arabidopsis* grown in the dark, *Arabidopsis* does not show a good response to earlier intermediates of brassinosteroid biosynthesis than cathasterone, such as campestanol and 6-oxocampestanol. This low response makes it difficult to investigate brassino-steroid biosynthesis upstream of cathasterone using feeding experiments even if the target site(s) of Brz exists.

Fujioka and Sakurai have demonstrated that there are at least two branched biochemical pathways to the end product, brassinolide: early C-6 oxidation or late C-6 oxidation pathway (7). In the late C-6 oxidation pathway, there are two hydroxylation steps of the side chain of campestanol leading to 6-deoxoteasterone via 6-deoxocathasterone. These steps are very similar to those in the early C-6 oxidation pathway. On the basis of these information it is not proper to rule out the possibility that Brz attacks these sites, as well as the step from cathasterone to teasterone. Further investigations are required to answer for these questions. Nevertheless, Brz exhibits its effect by reducing the supply of brassinolide in the plant.

Effect of Brz on Intact Plants and Cultured Cells

One advantage of Brz over brassinosteroid-deficient mutants in analyzing the functions of brassinosteroids in plants is its availability for application to a variety of plant species, at different growth stages, to tissues and cells, and to examine different biochemical reactions. Figure 4 shows the effect of Brz on *Arabidopsis*. In our experiments, Brz-treated plants showed morphological changes similar to those of brassinosteroids-deficient mutants of *Arabidopsis*:

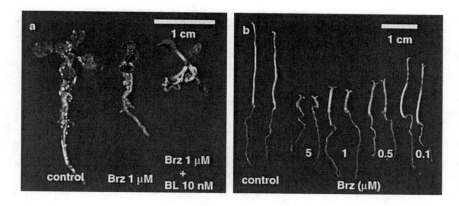

Figure 4. Morphology of Brz-treated and non-treated Arabidopsis. a: Arabidopsis seedlings (12-day-old) grown in the light. b: Arabidopsis seedlings (10-day-old) grown in the dark.

strong dwarfism with curly dark green leaves in the light (Figure 4a) and a de-etiolated phenotype with short hypocotyls and open cotyledons characteristic of light-grown plants when grown in the dark (Figure 4b). These results strongly suggested that Brz could be applied to other plant species to examine the function of brassinosteroids. As shown in Figure 5, Brz induced dwarfism with curly and dark green leaves in cress in the light (Figure 5a). In the dark, Brz induced photomorphogenetic changes to young seedlings of this plant. For example, when cress was treated with Brz, it showed short hypocotyls (Figure 5b), open cotyledons (Figure 5b) and the development of true leaves (Figure 5d) with the treatment of 1 µM of Brz. When cucumber was treated with Brz, it showed photomorphogenetic changes and rapid greening of cotyledons after irradiation for 3hr, conditions under which non-treated cucumber cotyledons retained their yellow color. Even though brassinosteroids are identified from monocotyledons, it generally shows less response to Brz in contrast to the results obtained in dicotyledons. This may reflect the different role of brassinosteroids between in monocotyledons and in dicotyledons.

Brz has also allowed the identification of interesting effects at the cellular level. In isolated mesophyll cells of *Zinnia elegans*, uniconazole prevented the differentiation of tracheary elements, and this inhibition was canceled by brassinolide application but not by GA. In this model system for xylem differentiation, Brz showed the same effect as uniconazole on the differentiation of tracheary elements (personal communication from Prof. H. Fukuda). This result implies that Brz can be used to clarify the involvement of endogenous brassinosteroids in specific steps in plant development.

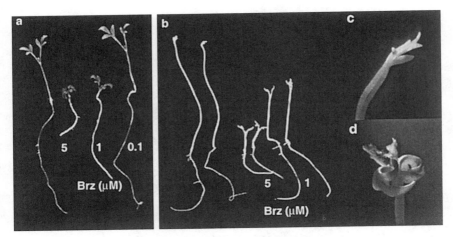

Figure 5. Morphology of Brz-treated and non-treated cress. a: Cress seedlings (10-day-old) grown in the light. b: Cress seedlings (8-day-old) grown in th dark. c and d: Cress seedlings (3-week-old) grown in the dark with or without 1 µM of Brz, respectively.

Conclusions

Brassinosteroid biosynthesis inhibitors are very useful for investigating brassinosteroid function in plants. However there are some limitations in the use of Brz. First, at a concentration of 10 µM or higher, the morphology of Brz-treated plants could not be restored completely to that of non-treated plants by brassinolide treatments possibly due to non-specific effect(s) on other aspects of plant metabolism. Second, plants that are larger than *Arabidopsis,* such as pea and cucumber, tend to require higher concentrations of Brz to bring about a change in morphology. This may be due to disturbance of uptake of Brz or its movement through the plant. Generally speaking, an ideal inhibitor for use with whole plants should possess the following characteristics: high specificity and affinity for the target enzyme, rapid uptake and transport within the plant, and a long-lasting effect. Intensive research on chemical modifications of lead compounds on the basis of chemical and biochemical knowledge should lead to the finding of an ideal inhibitor.

Finally, brassinosteroid biosynthesis inhibitors can be used to clarify the function of brassinosteroids in plants, as a complement to the use of brassino-steroid-deficient mutants. By analogy with the success in isolating new mutants using GA biosynthesis inhibitors, Brz may provide a ways of finding new brassinosteroid pathways, or isolation of mutants in brassinosteroid signal transduction. In addition to its use in basic science, Brz could also be developed as a new commercial plant growth regulator.

References

1 Mandava, N. *Annu. Rev. Plant Physiol. Plant Mol. Biol.* **1998**, *39*, 23-52.
2 Schlagnhaufer, C. D.; Arteca, R. N. *J. Plant Physiol.* **1991**, *138*, 191-194.
3 Iwasaki, T.; Shibaoka, H. *Plant Cell Physiol.* **1991**, *32*, 1007-1014.
4 Yamamoto, R.; Demura, T.; Fukuda, H. *Plant Cell Physiol.* **1997**, *38*, 980-983.
5 Azpiroz, R.; Wu Y.; LoCascio J. C.; Feldmann K. A. *The Plant Cell* **1998**, *10*, 219-230.
6 Clouse, S. D. *Plant J.* **1996**, *10*, 1-8.
7 Fujioka, S.; Sakurai, A. *Physiol. Plant.* **1997**, *100*, 710-715.
8 Yokota, T. *Trends in Plant Sci.* **1997**, *2*, 137-143.
9 Clouse, S. D.; Sasse, J. M. *Annu. Rev. Plant Physiol. Plant Mol. Biol.* **1998**, *49*, 427-451.
10 Feldmann, K.A.; Marks, M. D.; Christianson, M. L.; Quatrano, R. S. *Science* **1989**, *243*, 1351-1354.
11 Takahashi, T.; Gashc, A.; Nishizawa, N.; Chua, N. H. *Genes Dev.* **1995**, *9*, 97-107.
12 Kauschmann, A.; Jessop, A.; Koncz, C.; Szekeres, M.; Willmitzer, L.; Altmann, T. *Plant J.* **1996**, *9*, 701-713.
13 Szekeres, M.; Nemeth, K.; Koncz-Kalman, Z.; Mathur, J.; Kauschmann, A.; Altmann, T.; Redei, G. P.; Nagy, F.; Schell, J.; Koncz, C. *Cell* **1996**, *85*, 171-182.
14 Li, J.; Nagapal, P.; Vitart, V.; McMorris, T. C.; Chory, J. *Science* **1996**, *272*, 398-401.
15 Fujioka, S.; Li, J.; Choi, Y.; Seto, H.; Takatsuto, S.; Noguchi, T.; Watanabe, T.; Kuriyama, H.; Yokota, T.; Chory, J.; Sakurai, A. *The Plant Cell* **1997**, *9*, 1951-1962.
16 Nomura, T.; Nakayama, M.; Reid, J. B.; Takeuchi, Y.; Yokota, T. *Plant Physiol* **1997**, *113*, 31-37.
17 *Target Sites for Herbicide Action;* Böger, P.; Sandmann, G., eds.; CRC Press: Boca Raton, 1991; p 127.
18 *Gibberellins*; Takahashi, N.; Phinney, B. O.; MacMillan, J., eds.; Springer-Verlag: New York, 1991; p311-319.
19 Hartman, M-A.; Beneveniste, P. *Methods Enzymol.* **1987**, *148*, 632-650.
20 Burden, R.S.; Cooke, D.T.; Carter, G.A. *Phytochemistry* **1989**, *28*, 1791-1804.
21 Köller, W. *Physiol. Plant.* **1987**, *71*, 309-314.
22 Hartman, M-A. *Trends in Plant Sci.* **1998**, *3*, 170-175.
23 Choe, S.; Noguchi, T.; Fujioka, S.; Takatsuto, S.; Tissier, C. P.; Gregory, B. D.; Ross, A. S.; Tanaka, A.; Yoshida, S.; Tax, F. E.; Feldmann, K. A. *The Plant Cell* **1999**, *11*, 207-221.

24 *Target Sites in Fungicide Action*; Köller, W. ed.; CRC Press: Boca Raton, 1992; p 119-205.

25 Sakurai, A.; Fujioka, S. *Biosci. Biotech. Biochem.* **1997**, *61*, 757-762.

26 *Gibberellins*; Takahashi, N.; Phinney, B. O.; MacMillan, J., eds.; Springer-Verlag: New York, 1991; p339-349.

27 Takano, H.; Oguri, Y.; Kato, T. *J. Pestic. Sci.* **1986**, *11*, 373-378.

28 Takano, H.; Oguri, Y; Kato, T. *J. Pestic. Sci.* **1983**, *8*, 575-582.

29 Min, Y.K.; Asami, T.; Fujioka, S.; Murofushi, N.; Yamaguchi, I.; Yoshida, S. *Bioorg. Med. Chem. Lett.* **1999**, *9*, 425-430.

30 Yopp, J.; Mandava, N.; Sasse, J. M. *Physiol. Plant.* **1981**, *53*, 445-452.

31 Jones-Held, S.; VanDoren, M.; Lockwood, T. *J. Plant Growth Regul.* **1996**, *15*, 63-67.

32 Choe, S.; Dilkes, B. P.; Fujioka, S.; Takatsuto, S.; Sakurai, A.; Feldmann, K. A. *The Plant Cell* **1998**, *10*, 231-243.

Chapter 24

Electrophysiological Studies of Dopamine Release in Murine Striatal Slices and Its Role in the Neurotoxic Action of Cyclodienes

Ethan R. Freeborn and Jeffrey R. Bloomquist

Department of Entomology, Virginia Polytechnic Institute
and State University, Blacksburg, VA 24060

Cyclodiene mode of action has been attributed to two mechanisms: facilitation of neurotransmitter release and antagonism of the inhibitory neurotransmitter, γ-aminobutyric acid (GABA). In striatal brain slices, cyclodienes depressed neuronal firing, consistent with dopamine release from known striatal dopaminergic projections. In contrast, application of the GABA antagonist, picrotoxinin, produced excitation. The inhibitory action of dieldrin was blocked by the dopamine D_1 receptor antagonist, fluphenazine, verifying that released dopamine was responsible for inhibition of striatal neurons. These results suggest that cyclodiene-evoked neurotransmitter release may contribute significantly to the neurotoxicity of these compounds in the mammalian central nervous system.

Introduction

The GABA receptor/chloride channel complex is considered to be the primary site of cyclodiene action. A number of studies have shown that cyclodienes inhibit GABA-induced Cl^- influx (1-3) and the binding of $[^{35}S]tert$-butylbicyclophosphorothionate (TBPS), a known Cl^- channel ligand (4). Further, there is a good correlation between mammalian toxicity and inhibition of chloride uptake produced by the cyclodienes (1-4). Through this antagonism of inhibitory GABAergic pathways, central neuronal inhibition is reduced and an increased probability of hyperexcitation results. Such interference exacted upon central GABAergic neurons results in seizures, convulsions, and death.

Early studies on cyclodiene mode of action showed an effect on transmitter release. Dieldrin caused excitatory effects in cockroach abdominal ganglion preparations that were attributed to augmented release of acetylcholine (5). Similarly, aldrin-*trans*-diol, a metabolite of dieldrin (6), and lindane (7) were found to cause a rapid and marked increase in miniature frog motor end-plate frequency, an effect also apparently due to the release of presynaptic stores of acetylcholine. Heptachlor epoxide was later found to inhibit synaptic Ca^{2+}, Mg^{2+} - ATPases in rat brain synaptosomes (8). This inhibition was hypothesized to account for the ability of heptachlor epoxide to release preloaded $[^{14}C]$glutamate from isolated rat brain synaptosomes by increasing intracellular calcium concentration (9). Recent studies have shown a potent ability of cyclodienes to release dopamine from striatal synaptosomes, an effect that may be related to the development of environmentally-induced parkinsonism (10,11). Thus, we undertook an investigation to assess the relative contributions of transmitter release and GABA antagonism to the neurotoxic actions of these compounds.

Murine Striatal Slice Recordings

Brain slice electrophysiology was employed, in order to determine effects on a defined dopaminergic pathway, and to provide information on cyclodiene effects on intact nervous tissue, since previous studies of dopamine release had used synaptosomes. A technique for brain slice preparation was adapted from those of Brooks-Kayal *et al.* (12) and Fountain *et al.* (13). Mice were anesthetized with halothane, the brain rapidly removed, and 300 micron coronal slices (Figure 1) were cut with a vibrating tissue slicer.

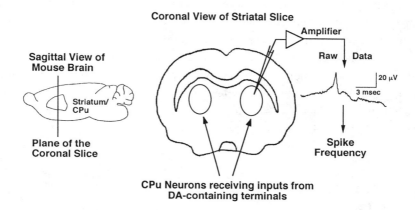

Figure 1 - Preparation of slices and experimental setup used for extracellular recording of neurons within the caudate putamen (CPu), which is part of the striatum. The CPu was identified visually using a mouse brain atlas (14).

The slices then were transferred to an incubation chamber and allowed to recover for a period of 3-4 hours. Extracellular recordings were made with glass micropipettes, and raw spike output was converted to a frequency, on line, for later statistical analysis.

Extracellular recordings of nerve activity were made from the indirect pathway, a subset of the projections from the substantia nigra that terminate within the striatum and serve to control the initiation and maintenance of movement (15). The indirect pathway is composed, in part, of an inhibitory dopaminergic innervation impinging upon GABAergic neurons that project from the striatum to the lateral globus pallidus. Activation of the presynaptic dopaminergic neurons in this circuit inhibits firing of target cells, and this inhibition is mediated by D_1 dopamine receptors (15). Because they are inhibted by dopamine, these cells make an ideal neural substrate for differentiating a synaptic effect of released dopamine from the hyperexcitation expected from GABA antagonism.

284

Dopamine Inhibition of Neuronal Discharge

Neurons of the indirect pathway were identified by their inhibitory response to exogenously applied dopamine. A dopamine concentration of 400 μM was found to consistently produce inhibition without causing permanent depression of firing activity. Neurons were found to exhibit an inhibitory response to the perfusion of dopamine within 1-3 minutes after the start of perfusion (Figure 2), and washout of the dopamine effect was typically incomplete. The firing frequency of indirect pathway neurons treated with dopamine was found to be significantly less than the baseline firing rates in the same neurons before treatment ($P < 0.05$, T-test).

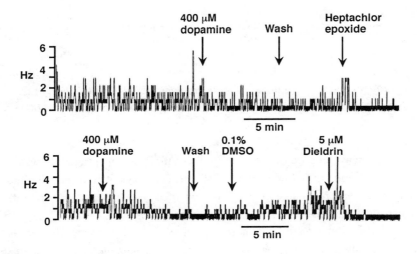

Figure 2. Heptachlor epoxide (5 μM) and dieldrin (5 μM) reduced the firing frequency of putative striatal indirect pathway neurons. The apparent increase in firing frequency at the time of the heptachlor epoxide application is an artifact of the drop-on treatment.

Cyclodienes Inhibit Nerve Discharge

The cyclodienes, dieldrin and heptachlor epoxide, caused a virtually complete cessation of firing within indirect pathway neurons at a concentration of 5 μM (Figure 2). These effects occurred from 15 seconds to 1 minute following the application of the insecticides. Washout periods extending to 15 minutes were ineffective in restoring pretreatment levels of firing activity. The solvent vehicle dimethyl sulfoxide (DMSO) was also applied to slices (0.1% solvent concentration) and found not to produce any significant change in firing frequency. Spike frequencies following 5 μM dieldrin or 5 μM heptachlor epoxide treatments were found to be significantly less than the firing frequencies just before insecticide treatment ($P < 0.05$, T-test). Due to the variability of the data collected in the heptachlor epoxide trials, a paired t-test comparison was made following a logarithmic transformation of the raw data. Concentration-response experiments with dieldrin found that 50% inhibition of nerve firing occurred at a concentration of 1.5 μM (IC_{50} value).

Picrotoxinin Increases Nerve Firing Rates

In contrast to the cyclodienes, application of 20 μM picrotoxinin (PTX) by perfusion was found to increase the firing frequency of indirect pathway neurons (Figure 3).

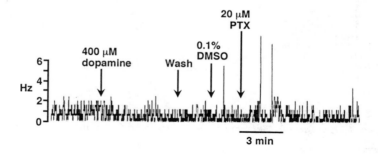

Figure 3. Exposure to PTX elevated the firing rates of striatal neurons.

This excitatory response was observed 1-2 minutes following application. At the peak levels of PTX-induced excitation, neural discharge rates were seen to increase an average of 240% above the pre-treatment firing rate. The duration of PTX-induced excitation was found to typically last for 1-2 minutes and there was no evidence of blockage. Unlike dieldrin-mediated inhibition, the excitatory effect of PTX was transient, and was not lengthened by continuous perfusion (>10 minutes). The mean frequencies of neural discharge measured before and after PTX treatment were found to be not quite significantly different ($P < 0.08$), but the direction of the effect was clearly opposite that of dieldrin. Due to the variability in the PTX trials, this paired t-test comparison was also made following a logarithmic transformation of the data.

Fluphenazine Blockage of Cyclodiene Inhibition

If the inhibitory action of cyclodienes upon indirect pathway neurons is mediated by released dopamine, a D_1 antagonist should reverse or prevent this effect. Accordingly, the cyclodiene depression of indirect pathway neuronal activity was blocked by a 5 minute pre-incubation with the dopamine antagonist, fluphenazine, at a concentration of 20 μM (Figure 4).

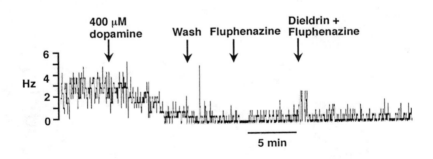

Figure 4. Exposure to dieldrin (5 µM) following a five minute perfusion of fluphenazine (20 µM), did not depress firing in indirect pathway neurons. The apparent increase in firing frequency at dieldrin application is an artifact of drop-on treatment.

In cells inhibited by dopamine, fluphenazine alone did not significantly alter the firing of striatal neurons during a 5 minute incubation. However, the expected inhibitory effect of a subsequent treatment with dieldrin was completely eliminated in the continued presence of flufenazine. A paired T-test comparison of the average firing rates immediately before (fluphenazine) and after the application of dieldrin (fluphenazine + dieldrin) indicated that the application of dieldrin had no significant effect on the firing rate (P = 0.32).

Significance of the Results

The indirect pathway neurons, such as those examined in this study, are known to be transiently inhibited by dopamine via D_1 receptors (15). The delay of the dopamine effect and the high concentrations of dopamine required to elicit a response may have resulted from the perfused dopamine having to penetrate into the slice, as well as being actively removed from the synapse by the dopamine reuptake transporter. Additionally, the effect of dopamine was difficult to reverse by washing. This observation may reflect the metabotropic character of the dopamine D_1 receptor, where elevation of second messengers in the neuron takes longer to subside than simple removal of agonist by washing.

Considering their well-documented action as GABA antagonists, the inhibitory effect on neuronal firing observed with dieldrin and heptachlor epoxide was inconsistent with this mechanism. We hypothesize that the inhibition of firing was due to a cyclodiene-evoked release of dopamine from nigral terminals and the subsequent postsynaptic D_1 reception of the neurotransmitter by dopamine-sensitive indirect pathway neurons. This conclusion is supported by the fact that cyclodienes mimicked the inhibitory action of exogenously applied dopamine and also by the ability of flufenazine to block the inhibitory effect of the cyclodienes. In contrast, application of picrotoxinin caused hyperexcitation of indirect pathway neurons. Since previous studies found that this compound does not release transmitters from preloaded synaptosomes (10), we attribute this action to GABA antagonism, and it indicates there were functional GABAergic pathways within the slice. At the concentrations used in this study (1-5 μM) no excitatory effects, indicative of GABA antagonism, were noted with the cyclodienes. Although cyclodiene concentrations should have been sufficient to elicit a GABA-mediated excitatory effect, this effect was evidently overwhelmed by an inhibitory action upon the neurons, mediated by dopamine release from presynaptic stores.

The relative contribution of these two mechanisms to the overall neurotoxicity of cyclodienes may be inferred from their relative potencies for antagonizing GABA compared to that for releasing transmitter. Dieldrin and heptachlor epoxide have been reported to inhibit GABA-gated chloride flux into mouse brain vesicles with EC_{50} values of 14 μM and 18 μM, respectively (3) and into rat brain microsacs with reported EC_{50} values of 3.3 μM and 0.45 μM, respectively (16). Whole-cell patch clamp recordings of rat dorsal root ganglion have shown chloride channel blockage by cyclodienes with EC_{50} values in the nanomolar range (17). In comparison, an IC_{50} value of 1.5 μM was found for dieldrin-induced inhibition of nerve firing that correlates nicely with a concentration of 1.4 μM that releases 50% of labeled dopamine from striatal synaptosomes (data not shown). Thus, measures of transmitter release are of similar potency to assays of GABA antagonism, suggesting that evoked neurotransmitter release may play an important role in cyclodiene neurotoxicity, *in vivo*.

A final question to be addressed is what relevance these findings may have for estimating any human health impacts resulting from cyclodiene exposure. The cellular degeneration and reductions in dopamine content of nigrostriatal projections observed in idiopathic Parkinson's disease (18) appear to have an environmental etiology (19). In addition, significantly higher brain levels of dieldrin are associated with an increased incidence of Parkinson's disease in humans (20). Lipophilic cyclodiene insecticides are sequestered in brain (20) and adipose tissue (21) and are present in the body at concentrations that may be sufficient to augment neurotransmitter release. The increased synaptic levels of dopamine and the subsequent upregulation of the presynaptic dopamine transporter we have observed (22) could result in neuronal stress, since high levels of dopamine are known to be neurotoxic (23). Alternatively, sensitization of dopaminergic neurons to other toxicants with affinity for the dopamine transporter could occur. Known dietary and metabolism-generated toxins with an affinity for the transporter include certain tetrahydroisoquinolines and β-carbolines that gain access to nerve terminals via the transporter and then inhibit mitochondrial respiration (24). Thus, increased dopamine release and upregulation of the dopamine transporter could synergize the actions of these other toxicants, and possibly play a role in pesticide-induced parkinsonism.

Acknowledgments

This material is based upon work supported by the Cooperative State Research Service, U. S. Department of Agriculture, under Hatch Project No. 6122040.

References

1. Abalis, I. M.; Eldefawi, M. E.; Eldefrawi, A. T. Effects of insecticides on GABA-induced chloride influx into rat brain vesicles. *J. Toxicol. Environ. Health* **1986,** *18,* 13-23.
2. Bloomquist, J. R.; Soderlund, D. M. Neurotoxic insecticides inhibit GABA-dependent chloride uptake by mouse brain vesicles. *Biochem Biophys. Res. Commun.* **1985,** *133,* 37-43.
3. Bloomquist, J. R.; Adams, P. M.; Soderlund, D. M. Inhibition of γ-aminobutyric acid-stimulated chloride flux in mouse brain vesicles by polychlorocycloalkane and pyrethroid insecticides. *Neurotoxicol.* **1986,** *7,* 11-20.
4. Cole, L. M.; Casida, J. E. Polychlorocycloalkane insecticide-induced convulsions in mice in relation to disruption of the GABA-regulated chloride ionophore. *Life Sci.* **1986,** *39,* 1855-1862.
5. Shankland, D.L.; Schroeder, M. E. Pharmacological evidence for a discrete neurotoxic action of dieldrin (HEOD) in the American cockroach, Periplaneta americana (L.). *Pestic. Biochem. Physiol* **1973,** *3,* 77-86.
6. Akkermans, L. M.; van den Bercken, J.; van der Zalm J. M.; Straaten, H. W. Effects of dieldrin (HEOD) and some of its metabolites on synaptic transmission in the frog motor end-plate. *Pest. Biochem. Physiol.* **1974,** *4,* 313-324.
7. Publicover, S. J.; Duncan, C. J. The action of lindane in accelerating the spontaneous release of transmitter at the frog neuromuscular junction. *Naunyn-Schmiedeberg's Arch. Pharmacol.* **1979,** *381,* 179-182.

8. Yamaguchi, I.; Matsumura, F.; Kadous, A. Inhibition of synaptic ATPases by heptachlor epoxide in rat brain. *Pest. Biochem. Physiol.* **1979**, *11*, 285-293.

9. Yamaguchi, I.; Matsumura, F.; Kadous, A. Heptachlor epoxide: effects on calcium-mediated transmitter release from brain synaptosomes in rat. *Biochem. Pharmacol.* **1980**, *29*, 1815-1823.

10. Kirby M. L.; Bloomquist J. R. Neurotoxicity of the organochlorine insecticide heptachlor and its role in Parkinsonism. *Fund. Appl. Toxicol. Suppl. The Toxicologist* **1997**, *36*, 343.

11. Bloomquist, J. R.; Kirby, M. L.; Castagnoli, K.; Miller, G. W. *Effects of Heptachlor Exposure on Neurochemical Biomarkers of Parkinsonism.* In: Progress in Neuropharmacology and Neurotoxicology of Pesticides and Drugs; Beadle D. J., Ed.; Society of Chemical Industry/Royal Society of Chemistry, Cambridge, United Kingdom, 1999, pp. 195-203.

12. Brooks-Kayal, A. R.; Shumate, M. D.; Jin, H.; Rikhter, T. Y.; Coulter, D. A. Selective changes in single cell $GABA_A$ receptor subunit expression and function in temporal lobe epilepsy. *Nat. Med.* **1998**, *10*, 1166-72.

13. Fountain, S. B.; Ting, Y. L.; Teyler, T. J. The *in vitro* hippocampal slice preparation as a screen for neurotoxicity. *Toxicol. in vitro* **1990**, *6*, 77-87.

14. Franklin, K.; Paxinos, G. *The Mouse Brain in Stereotaxic Coordinates;* Academic Press: San Diego, California, 1997; figures 25-37.

15. Purves, D.; Augusting, G. J.; Fitzpatrick, D.; Katz, L. C.; LaMantia, A-S.; McNamara, J. O. *Neuroscience.* Sinaur Associates, Inc. Sunderland, MA. 1997; pp. 329-344.

16. Gant, D. B.; Eldefrawi, M. E.; Eldefrawi, A. T. Cyclodiene insecticides inhibit $GABA_A$ receptor-mediated chloride transport. *Tox. Appl. Pharmacol.* **1987**, *88*, 313-321.

17. Nagata, K.; Narahashi, T. Dual action of the cyclodiene insecticide dieldrin on the GABA receptor-chloride channel complex of rat dorsal root ganglion neurons. *J. Pharmacol. Exp. Ther.* **1993**, *269*, 164-170.

18. Bowman, W. C.; Rand, M. J. *Textbook of Pharmacology*, 2nd edition. Blackwell Scientific Publications, Oxford, 1980; pp. 18.17-18.25.

19. Tanner, C,; Ottman, R.; Goldman, S.; Ellenberg, J.; Chan, P.; Mayeux, R.; Langston, J. Parkinson disease in twins: an etiologic study. *JAMA.* **1999**, *281*, 341-346.

20. Fleming, L.; Mann, J. B.; Bean, J.; Briggle, T.; Sanchez-Ramos, J. R. Parkinson's disease and brain levels of organochlorine insecticides. *Ann. Neurol.* **1994,** *36,* 100-103.

21. Burgaz, S.; Afkham, B. L.; Karakaya, A. E. Organochlorine pesticide contaminants in human adipose tissue collected in Ankara (Turkey) 1991-1992. *Bulletin Environ. Contam. Toxicol.* **1994,** *53,* 501-508.

22. Miller, G. W.; Kirby, M. L.; Levey, A. I.; Bloomquist, J. R. Heptachlor Alters Expression and Function of Dopamine Transporters. *Neurotoxicol.* **1999,** *20,* 631-638.

23. Filloux, F.; Townsend, J. Pre- and postsynaptic neurotoxic effects of dopamine demonstrated by intrastriatal injection. *Exp. Neurol.* **1993,** *119,* 79-88.

24. Castagnoli, N.; Rimoldi, J.; Bloomquist, J.; Castagnoli, K. Potential metabolic bioactivation pathways involving cyclic tertiary amines and azaarenes. *Chem. Res. Toxicol.* **1997,** *10,* 924-940.

Author Index

Subject Index

302